W9-ADN-616

London Mathematical Society Lecture Note Series. 10

Numerical Ranges II

F.F. BONSALL and J. DUNCAN

Cambridge · At the University Press · 1973

Published by the Syndics of the Cambridge University Press
Bentley House, 200 Euston Road, London NW1 2DB
American Branch: 32 East 57th Street, New York, N. Y. 10022

© Cambridge University Press 1973

Library of Congress Catalogue Card Number: 71-128498

ISBN: 0 521 20227 2

Printed in Great Britain
at the University Printing House, Cambridge
(Brooke Crutchley, University Printer)

1362179

Contents

MATH.-SCI.
QA
322.2
.B66

Introduction

This volume is a sequel to Numerical ranges of operators on normed spaces and of elements of normed algebras, which is here denoted by NRI. Although it appeared in 1971, NRI was written in 1969, and since then the subject has made vigorous progress, reaching a high point with the Conference on Numerical Ranges, held in Aberdeen, July 1971. This conference gave us an unusually good opportunity to see the scope of the subject, which is much less specialized than the title might suggest.

A comparison of the present volume with NRI will show that the theory of numerical ranges has become immensely richer both in depth and in width. The contents have been grouped into three chapters: 5. Spatial numerical ranges; 6. Algebra numerical ranges; 7. Further ranges.

In Chapter 5 we are mainly concerned with the improvement of the Bishop-Phelps theorem due to Bollobás [115] and with applications of this useful tool. We also give the remarkable theorem of Zenger [78] on the inclusion of the convex hull of the point spectrum in the spatial numerical range $V(T)$, and the equally remarkable results of Crabb [136] and Sinclair [202] concerning points of $Sp(T) \cap \partial V(T)$.

NRI contained an inequality (Theorem 4. 8) relating the norms of iterates to the numerical radius and a remark that this inequality had been proved to be best possible in a strong sense. Chapter 6 contains a systematic approach to such best possible inequalities through the theory of the extremal algebra $Ea(K)$ which has been developed by Bollobás [117] and Crabb, Duncan, and McGregor [134]. Here the numerical range comes into contact with interesting function theoretic ideas. Chapter 6 also contains an account of the striking progress made in the study of Hermitian elements and related concepts by Berkson [109], Browder [125], Berkson, Dowson and Elliott [113], Moore [186], Sinclair [204], and others. The proof of the Vidav-Palmer theorem, which bulked large in NRI, receives some final touches.

In our final chapter we give a brief survey of essential numerical ranges, joint numerical ranges, and matrix ranges, and end with an axiomatic approach to numerical ranges. The theory of the essential numerical range has been developed with force by Fillmore, Stampfli and Williams [151], and by Anderson [100] who has established the important operator theoretic significance of the condition $0 \in$ Wess(T). Two concepts of matrix range have been developed, the analogue of the algebra numerical range by Arveson [104], and the analogue of the spatial numerical range by S. K. Parrott (unpublished), and both concepts have been shown to provide complete sets of unitary invariants for certain wide classes of compact operators.

We are very much aware that our account of numerical ranges remains unbalanced in that we have not attempted to give an account of the applications to initial value problems. When we came to study the literature of this important subject, we soon concluded that we were not qualified to do it justice, and we hope that some expert in the field will fill this gap. A valuable bibliography is given in Calvert and Gustafson [129]. We have also refrained from developing numerical ranges in real algebras; significant advances in this area may be found in Lumer [176] and McGregor [180]. The bibliography in NRI lacked an adequate coverage of numerical ranges for Hilbert space operators and we have attempted to repair this deficiency in the present volume. The present bibliography also contains several other items which are not mentioned in the text.

This volume being a companion to NRI, we have continued the same mode of references. To simplify back references we have numbered the sections in NRII starting with §15. Likewise the bibliography in NRII starts with [100], so that [n] with $n < 100$ refers to the bibliography in NRI.

Many authors have given us valuable help by making their work available to us before publication, and we wish to acknowledge particularly the help received from W. B. Arveson, E. Berkson, B. Bollobás, A. Browder, M. J. Crabb, H. R. Dowson, K. Gustafson, L. A. Harris, C. M. McGregor, R. T. Moore, T. W. Palmer, S. K. Parrott, A. M. Sinclair, and J. P. Williams. As in NRI only a very few results appear in print here for the first time. Most of the material in this volume has been the subject of seminar talks by the authors, and has benefitted from the resulting criticism. We have had many valuable conversations with G. A. Johnson which have left their mark particularly on §§36, 37. In the elimination of errors we have been greatly helped by D. J. Baker and A. W. Tullo who have read the manuscript.

The whole manuscript has been most expertly typed by Miss Christine Bourke.

January 1972

University of Edinburgh University of Stirling

5 · Spatial numerical ranges

15. SOME ELEMENTARY OBSERVATIONS ON THE SPATIAL NUMERICAL RANGE

Let X denote a normed linear space over $\mathbb{C}$, X' its dual space, $S(X)$ its unit sphere, $\Pi(X)$ the subset of $X \times X'$ defined by

$$\Pi(X) = \{(x, f) \in S(X) \times S(X') : f(x) = 1\}.$$

Given $x \in S(X)$, let $D(X, x) = \{f \in S(X') : f(x) = 1\}$.

We recall that the spatial numerical range $V(T)$, for $T \in B(X)$, is defined by $V(T) = \{f(Tx) : (x, f) \in \Pi(X)\}$. Given $x \in S(X)$, let

$$V(T, x) = \{f(Tx) : f \in D(X, x)\}.$$

Clearly $V(T) = \bigcup\{V(T, x) : x \in S(X)\}$.

We collect a few elementary observations on $V(T)$, $V(T, x)$ and $\Pi(X)$ into this section. The first is derived from a remark of J. P. Williams on the algebra numerical range.

Lemma 1. Let $x \in S(X)$, $T \in B(X)$. Then $V(T, x)$ is the set of all complex numbers λ such that

$$|\lambda - \zeta| \leq \|(T - \zeta I)x\| \qquad (\zeta \in \mathbb{C}). \tag{1}$$

Proof. If $\lambda \in V(T, x)$, there exists $f \in D(X, x)$ with $f(Tx) = \lambda$, and so

$$|\lambda - \zeta| = |f((T - \zeta I)x)| \leq \|(T - \zeta I)x\| \qquad (\zeta \in \mathbb{C}).$$

Suppose on the other hand that λ satisfies the inequalities (1). If $Tx \in \mathbb{C}x$, we take $\zeta \in \mathbb{C}$ with $Tx = \zeta x$, i. e. with $(T - \zeta I)x = 0$. Then (1) gives $\lambda = \zeta = f(Tx)$ for arbitrary $f \in D(X, x)$. Suppose then that x, Tx are linearly independent and define f_0 on their linear span by

$$f_0(\alpha Tx + \beta x) = \alpha\lambda + \beta \qquad (\alpha, \beta \in \underset{\sim}{C}).$$

The inequalities (1) imply that $\|f_0\| \le 1$, and we have $f_0(x) = 1$, $f_0(Tx)=\lambda$. The Hahn-Banach theorem now gives an extension f of f_0 with $f \in D(X, x)$ and $f(Tx) = \lambda$.

Remark. This lemma exhibits $V(T, x)$ as an intersection of closed discs.

Lemma 2. <u>Let</u> X_1 <u>be a non-zero linear subspace of</u> X, <u>let</u> $T \in B(X)$, <u>and let</u> $TX_1 \subset X_1$. <u>Then</u>

(i) $V(T|_{X_1}) \subset V(T)$,

(ii) $V(T|_{X_1}, x_1) = V(T, x_1) \qquad (x_1 \in S(X_1))$.

Proof. (i) follows from (ii) which is an immediate corollary of Lemma 1.

Remarks. (1) If X_1 is a closed linear subspace of X and $X_1 \neq X$, then the difference space $Y = X - X_1$ is a non-zero normed space with respect to the canonical norm $\|y\| = \inf\{\|x\| : x \in y\}$ $(y \in Y)$. Given $T \in B(X)$ with $TX_1 \subset X_1$, we obtain an operator $U \in B(Y)$ given by $Uy = Tx + X_1$ $(x \in y \in Y)$, and it is natural to ask whether $V(U) \subset V(T)$. Using rather deeper arguments, we prove in Theorem 17.5 that if X is complete then $V(U) \subset V(T)^-$.

(2) Let X_1, X_2 be non-zero subspaces of X such that $X = X_1 \oplus X_2$, and suppose that

$$\|x_1 + x_2\| = \|x_1\| + \|x_2\| \qquad (x_1 \in X_1,\ x_2 \in X_2).$$

Let $T \in B(X)$ with $TX_1 \subset X_1$, $TX_2 \subset X_2$. Then

$$V(T) = \{\alpha\lambda + (1-\alpha)\mu : \alpha \in [0,1],\ \lambda \in V(T|_{X_1}),\ \mu \in V(T|_{X_2})\}.$$

The same conclusion holds if, for some p with $1 \le p \le \infty$, we have

$$\|x_1 + x_2\| = (\|x_1\|^p + \|x_2\|^p)^{1/p} \qquad (x_1 \in X_1,\ x_2 \in X_2).$$

Lemma 3. Let X_R denote the space X regarded as a normed linear space over $\underset{\sim}{R}$. Then the mapping $f \to \mathrm{Re}\, f$ is an isometric real linear mapping of X' onto X_R'.

Proof. Given $f \in X'$, it is clear that $\mathrm{Re}\, f \in X_R'$ and that $\|\mathrm{Re}\, f\| \le \|f\|$. Given $\zeta \in \underset{\sim}{C}$ with $|\zeta| = 1$, we have

$$|\mathrm{Re}(\zeta f(x))| = |\mathrm{Re}\, f(\zeta x)| \le \|\mathrm{Re}\, f\|\ \|\zeta x\| = \|\mathrm{Re}\, f\| . \|x\|.$$

Since we may choose such a complex number ζ with $\mathrm{Re}(\zeta f(x)) = |f(x)|$, this gives $|f(x)| \le \|\mathrm{Re}\, f\| . \|x\|$, and so $\|f\| = \|\mathrm{Re}\, f\|$.

It is clear that the mapping $f \to \mathrm{Re}\, f$ is a real linear mapping, and so it only remains to prove that the mapping is surjective. Given $g \in X_R'$, define f by

$$f(x) = g(x) - ig(ix) \qquad (x \in X).$$

Then $f \in X'$, and $\mathrm{Re}\, f = g$.

Corollary 4. The mapping $(x, f) \to (x, \mathrm{Re}\, f)$ maps $\Pi(X)$ onto $\Pi(X_R)$.

Proof. Given $(x, f) \in \Pi(X)$, we have $\|\mathrm{Re}\, f\| = 1$ and $(\mathrm{Re}\, f)(x) = 1$, so $(x, \mathrm{Re}\, f) \in \Pi(X_R)$. Given $(x, g) \in \Pi(X_R)$, there exists $f \in S(X')$ with $\mathrm{Re}\, f = g$. Since $|f(x)| \le 1$ and $(\mathrm{Re}\, f)(x) = 1$, we have $f(x) = 1$ and so $(x, f) \in \Pi(X)$.

Corollary 5. Let $T \in B(X)$, and let T_R denote T regarded as an operator on X_R. Then

$$V(T_R) = \mathrm{Re}\, V(T).$$

Proof. Immediate from Corollary 4.

The following results on upper semi-continuous set valued mappings will be useful for §20.

Definition 6. Let E, F be topological spaces, and let 2^F denote the set of all subsets of F. A mapping $\phi : E \to 2^F$ is said to be upper semi-continuous (u. s. c.) if for each $x \in E$ and each neighbourhood U of $\phi(x)$, there exists a neighbourhood V of x such that

$$y \in V \Rightarrow \phi(y) \subset U.$$

Roughly, if y is near x then all points of $\phi(y)$ are near $\phi(x)$.

For metric spaces E, F with F compact there is a convenient 'closed graph' criterion for upper semi-continuity, as follows.

Lemma 7. Let E, F be metric spaces with F compact, let ϕ be a mapping of E into 2^F such that $\phi(x)$ is closed for each $x \in E$. Then ϕ is u. s. c. if and only if

$$x_n \in E,\ y_n \in \phi(x_n)\ (n=1, 2, \ldots),\ x = \lim_{n\to\infty} x_n,\ y = \lim_{n\to\infty} y_n \Rightarrow y \in \phi(x).$$

Proof. Assume that ϕ is u. s. c., that $x_n \in E$, $y_n \in \phi(x_n)$, $x = \lim_{n\to\infty} x_n$ and $y = \lim_{n\to\infty} y_n$. Let U be a closed neighbourhood of $\phi(x)$. Then there exists a neighbourhood V of x such that $\phi(z) \subset U$ $(z \in V)$. Since $x_n \in V$ for all sufficiently large n, we have $y_n \in \phi(x_n) \subset U$ for all sufficiently large n, and so $y \in U$. Thus y belongs to every closed neighbourhood of $\phi(x)$, and since $\phi(x)$ is closed this gives $y \in \phi(x)$.

Assume next that ϕ satisfies the closed graph criterion, and that U is a neighbourhood of $\phi(x)$. If there is no neighbourhood V of x satisfying $\phi(z) \subset U$ $(z \in V)$, then for every positive integer n, there exists $x_n \in E$ such that $d(x_n, x) < \frac{1}{n}$ but $\phi(x_n) \not\subset U$. Choose $y_n \in \phi(x_n) \backslash U$. Since F is compact there exists a subsequence $\{y_{n_k}\}$ with $\lim_{k\to\infty} y_{n_k} = y \in F$. Since also $\lim_{k\to\infty} x_{n_k} = x$, we conclude that $y \in \phi(x)$. Since U is a neighbourhood of $\phi(x)$ we obtain $y_{n_k} \in U$ for all sufficiently large k, a contradiction.

Lemma 8. The mapping $x \to V(T, x)$ is an upper semi-continuous mapping of S(X) with the norm topology into the non-void compact convex subsets of $\underset{\sim}{C}$.

Proof. The sets $V(T, x)$ are non-void compact convex subsets of a compact disc in $\underset{\sim}{C}$, so an application of Lemma 7 will complete the proof.

Let $x_n \in S(X)$, $\lambda_n \in V(T, x_n)$, $\lim_{n\to\infty} \|x_n - x\| = 0$, $\lim_{n\to\infty} |\lambda_n - \lambda| = 0$. There exist $f_n \in D(x_n)$ with $\lambda_n = f_n(Tx_n)$. By the weak* compactness of the unit ball in X', there exists a weak* cluster point f of $\{f_n\}$ with $\|f\| \le 1$. Also

$$
\begin{aligned}
|1 - f(x)| &\le |f_n(x_n) - f_n(x)| + |f_n(x) - f(x)| \\
&\le \|x_n - x\| + |(f_n - f)(x)|,
\end{aligned}
$$

from which $f(x) = 1$, and so $f \in D(x)$. Finally,

$$
\begin{aligned}
|\lambda - f(Tx)| &\le |\lambda - \lambda_n| + |f_n(Tx_n) - f_n(Tx)| + |f_n(Tx) - f(Tx)| \\
&\le |\lambda - \lambda_n| + \|T\| \, \|x_n - x\| + |(f_n - f)(Tx)|,
\end{aligned}
$$

which gives $\lambda = f(Tx) \in V(T, x)$.

We end this section with a proof of the Toeplitz-Hausdorff theorem on the convexity of the spatial numerical range of an operator on a Hilbert space. The present proof is derived from the review by Halmos [157] of a proof due to Gustafson [155]. In this review Halmos outlines a proof derived from the work of Dekker [138] on joint numerical ranges. The present proof is a modification of this using ideas that are to be found in the proof in Halmos [30], Problem 166. In what follows, H is a Hilbert space with scalar product $(\, , \,)$, and as is usual in this context the spatial numerical range of T is denoted by $W(T)$; of course we have

$$W(T) = \{(Tx, x) : x \in S(H)\}.$$

Lemma 9. Let L be a self-adjoint element of $B(H)$, and let

$$E = \{x \in S(H) : (Lx, x) = 0\}.$$

Then E is arcwise connected.

Proof. Note first that if $x \in E$, then $e^{i\theta}x \in E$ $(\theta \in \underset{\sim}{R})$, and $e^{i\theta}x$

is joined to x by an arc in E. Assume then that a, b are linearly independent elements of E, choose $\theta \in \underset{\sim}{R}$ such that $e^{i\theta}(La, b) \in i\underset{\sim}{R}$ and take $c = e^{i\theta}a$. Then we have $(Lc, b) \in i\underset{\sim}{R}$, and it is enough to show that we can join c to b by an arc in E. With $0 \le \alpha \le 1$, let $x(\alpha) = (1-\alpha)c + \alpha b$. Then

$$(Lx(\alpha), x(\alpha)) = (1-\alpha)^2(Lc, c) + (1-\alpha)\alpha\{(Lc, b)+(Lb, c)\} + \alpha^2(Lb, b),$$

and $(Lc, b) + (Lb, c) = 2\,\mathrm{Re}(Lc, b) = 0$. Thus $(Lx(\alpha), x(\alpha)) = 0$, and we obtain the required arc by taking $u(\alpha) = \|x(\alpha)\|^{-1}\, x(\alpha)$.

Lemma 10. Let A, B be self-adjoint elements of B(H), and let

$$W = \{((Ax, x), (Bx, x)) : x \in S(H)\}.$$

Then W is a convex subset of $\underset{\sim}{R}^2$.

Proof. It is enough to prove that $W \cap l$ is connected when l is a straight line in $\underset{\sim}{R}^2$. Let l have equation $\alpha\xi + \beta\eta + \gamma = 0$, and let $L = \alpha A + \beta B + \gamma I$. The mapping π given by

$$\pi x = ((Ax, x), (Bx, x))$$

is continuous, and

$$E = \{x \in S(H) : (Lx, x) = 0\} = \{x \in S(H) : \pi x \in l\}.$$

Therefore $W \cap l = \pi E$ and is a connected set, by Lemma 9.

Theorem 11. (Toeplitz-Hausdorff.) Let $T \in B(H)$. Then W(T) is convex.

Proof. We have $T = A + iB$ with A, B self-adjoint; and with W as in Lemma 10, $W(T) = \{\xi + i\eta : (\xi, \eta) \in W\}$.

16. THE BISHOP-PHELPS-BOLLOBÁS THEOREM

Given a non-reflexive Banach space X there is a certain lack of symmetry in the relationship between the sets S(X), S(X'), Π(X). For, given $x \in S(X)$ there always exists $f \in S(X')$ such that $(x, f) \in \Pi(X)$; but, given $f \in S(X')$ there need not exist any $x \in S(X)$ such that $(x, f) \in \Pi(X)$. Let us call $f \in S(X')$ a _support functional_ if $(x, f) \in \Pi(X)$ for some x. The theorem of Bishop and Phelps [9] states that, when X is a Banach space, the set of support functionals is norm dense in $S(X')$, and we have seen already (NRI Theorem 9.4 and Theorem 10.1) that this has important implications for numerical ranges. B. Bollobás [115] has proved a stronger form of the Bishop-Phelps theorem which is even more significant for our subject, as follows.

Theorem 1. (Bishop-Phelps-Bollobás.) _Let_ X _be a Banach space, and let_ $0 < \varepsilon < 1$. _Given_ $z \in X$, $h \in S(X')$ _with_ $\|z\| \le 1$ _and_

$$|1 - h(z)| < \varepsilon^2/4,$$

there exists $(y, g) \in \Pi(X)$ _such that_ $\|y - z\| < \varepsilon$, $\|g - h\| < \varepsilon$.

Remarks. (1) Roughly speaking the theorem says that elements of $X \times X'$ that nearly satisfy the defining conditions for $\Pi(X)$ are close to elements of $\Pi(X)$ (in the product of the norm topologies).

(2) The Bishop-Phelps theorem is an immediate corollary. For, given $h \in S(X')$ we may choose $z \in X$ with $\|z\| \le 1$ and $|1 - h(z)| < \varepsilon^2/4$. Then, by Theorem 1, we have a support functional $g \in X'$ with $\|g - h\| < \varepsilon$.

The proof of Theorem 1 involves only minor changes from the proof of the Bishop-Phelps theorem [9]. Note that Lemma 15.3 reduces the proof of Theorem 1 to the case when the scalar field is $\underset{\sim}{R}$. For if X is over $\underset{\sim}{C}$, and $|1 - h(z)| < \varepsilon^2/4$, then

$$|1 - \mathrm{Re}\, h(z)| < \varepsilon^2/4.$$

Therefore if Theorem 1 has been proved for real scalars, there exist $g \in X_R'$ with $\|g\| = 1$ and $y \in S(X)$ with $g(y) = 1$ such that $\|y - z\| < \varepsilon$ and $\|g - \mathrm{Re}\, h\| < \varepsilon$. By Lemma 15.3, there exists $f \in X'$ with $g = \mathrm{Re}\, f$, and we have $\|f\| = 1$, $\|f-h\| = \|g-\mathrm{Re}\, h\| < \varepsilon$. Also

Re $f(y) = 1$, and $|f(y)| \leq 1$, from which Im $f(y) = 0$, and $f(y) = 1$.

Notation. We suppose for the rest of this section that X is a Banach space over $\underset{\sim}{R}$ and denote by U the closed unit ball of X.

Lemma 2. Let $f, g \in S(X')$, $\varepsilon > 0$, and $T = \{x \in \frac{2}{\varepsilon}U : f(x) = 0\}$. If $|g(x)| \leq 1$ $(x \in T)$, then either $\|f-g\| \leq \varepsilon$ or $\|f+g\| \leq \varepsilon$.

Proof. Suppose that $|g(x)| \leq 1$ $(x \in T)$, and let $X_0 = \{x \in X : f(x) = 0\}$. Then $X_0 \cap U = \frac{\varepsilon}{2}T$, and so

$$\sup\{|g(x)| : x \in X_0 \cap U\} \leq \frac{\varepsilon}{2}.$$

By the Hahn-Banach theorem $g|_{X_0}$ can be extended to the whole of X without change of norm; i.e. there exists $h \in X'$ with $h|_{X_0} = g|_{X_0}$ and $\|h\| \leq \frac{\varepsilon}{2}$. Since $(g - h)(X_0) = \{0\}$, there exists $\alpha \in \underset{\sim}{R}$ with $g - h = \alpha f$, i.e.

$$g - \alpha f = h. \tag{1}$$

Suppose first that $0 \leq \alpha \leq 1$. Then

$$\begin{aligned} \|g - f\| &\leq \|g - \alpha f\| + \|(\alpha - 1)f\| \\ &= \|g - \alpha f\| + 1 - \alpha \\ &= \|g - \alpha f\| + \|g\| - \|\alpha f\| \\ &\leq 2\|g - \alpha f\| = 2\|h\| \leq \varepsilon. \end{aligned}$$

Next suppose that $\alpha > 1$. Then $0 < \frac{1}{\alpha} < 1$, and (1) takes the form

$$f - \frac{1}{\alpha}g = -\frac{1}{\alpha}h. \tag{2}$$

Comparing (1) and (2), we now have

$$\|f - g\| \leq 2\|-\frac{1}{\alpha}h\| \leq \varepsilon.$$

We have now proved that $\|g - f\| \leq \varepsilon$ if $\alpha \geq 0$. Finally if $\alpha < 0$, we consider (1) in the form

$$g - (-\alpha)(-f) = h,$$

and conclude that $\|g - (-f)\| \le \varepsilon$.

Proof of Theorem 1. We have seen already that it is enough to suppose that X is a Banach space over $\underset{\sim}{R}$.

Let $z \in U$, $h \in S(X')$, $0 < \varepsilon < 1$, $|1 - h(z)| < \varepsilon^2/4$; and take $T = \{x \in \frac{2}{\varepsilon}U : h(x) = 0\}$.

We obviously have $h(z) > 0$. Let $\kappa = \frac{1}{h(z)}(1 + \frac{2}{\varepsilon})$, and define a relation $\le$ of partial order on U by:

$$x \le y \Longleftrightarrow \|x - y\| \le \kappa h(y - x).$$

Note that $x \le y \Rightarrow h(x) \le h(y)$, since $\kappa > 0$. Let $Z = \{x \in U : z \le x\}$. We prove that Z has a maximal element y.

Given a chain (totally ordered set) $W \subset Z$, $\{h(w)\}_{w \in W}$ is an increasing net in $\underset{\sim}{R}$ bounded above by 1. Therefore $\{h(w)\}_{w \in W}$ converges. Since

$$w \le w' \Rightarrow \|w' - w\| \le \kappa(h(w') - h(w)),$$

W is a Cauchy net in the Banach space X and therefore converges to $v \in U$. By continuity of h and the norm, we have $z \le w \le v$ $(w \in W)$, and so $v \in Z$; v is an upper bound for W in Z. Therefore Z is inductively ordered, and, by Zorn's lemma, it has a maximal element y.

Since $y \in Z$, we have $z \le y$, i. e.

$$\|y - z\| \le \kappa(h(y) - h(z)).$$

Since $y \in U$, we have $h(y) \le 1$, and so

$$\begin{aligned} \|y - z\| &\le (1 + \tfrac{2}{\varepsilon})\frac{1}{h(z)}(1 - h(z)) \\ &\le (1 + \tfrac{2}{\varepsilon})(\tfrac{\varepsilon}{2})^2(1 - (\tfrac{\varepsilon}{2})^2)^{-1} \\ &= \tfrac{\varepsilon}{2}(1 - \tfrac{\varepsilon}{2})^{-1} < \varepsilon. \end{aligned} \tag{3}$$

Let $C = co(U \cup T)$, and let p denote the Minkowski functional of C; i. e.

$$p(x) = \inf\{\alpha > 0 : \frac{1}{\alpha}x \in C\}.$$

Since $U \subset C$, we have

$$p(x) \le \|x\| \qquad (x \in X).$$

We prove that $p(y) = 1$. If not, then $p(y) < 1$, since $y \in U$, and so there exists a real number α with $0 < \alpha < 1$ and $\frac{1}{\alpha} y \in C$. Therefore, U and T being convex sets, we have

$$\frac{1}{\alpha} y = \lambda u + (1 - \lambda)t$$

for some $\lambda \in [0,1]$, $u \in U$, $t \in T$. Then since $h(z) > 0$,

$$h(z) \le h(y) = \alpha\lambda h(u) < h(u), \tag{4}$$

$$h(u - y) = (1 - \alpha\lambda)h(u) \ge (1 - \alpha\lambda)h(z). \tag{5}$$

Also $u - y = (1 - \alpha\lambda)u - \alpha(1 - \lambda)t$, and so

$$\begin{aligned}\|u - y\| &\le (1 - \alpha\lambda) + \alpha(1 - \lambda)\ \|t\| \\ &\le (1 - \alpha\lambda) + \alpha(1 - \lambda)\frac{2}{\varepsilon} \le (1 - \alpha\lambda)(1 + \frac{2}{\varepsilon}).\end{aligned} \tag{6}$$

From (5) and (6)

$$\|u - y\| \le \frac{1}{h(z)}(1 + \frac{2}{\varepsilon})\ h(u - y) = \kappa h(u - y).$$

Therefore $y \le u$. Since y is maximal, this gives $y = u$, which contradicts (4) and proves that $p(y) = 1$.

By the Hahn-Banach theorem, there exists a linear functional g on X with $g \le p$ and $g(y) = p(y) = 1$. Since $y \in U$ and $p(x) \le \|x\|$ $(x \in X)$, we have $\|y\| = 1$. Also $g \in X'$ and $\|g\| \le 1$. Therefore

$$(y, g) \in \Pi(X).$$

By (3), $\|y - z\| < \varepsilon$.

We have $h, g \in S(X')$, and, since $T \subset C$,

$$g(x) \le p(x) \le 1 \qquad (x \in T).$$

Also $-T = T$, and so

$$|g(x)| \leq 1 \qquad (x \in T).$$

Therefore, by Lemma 2, either $\|g - h\| \leq \varepsilon$ or $\|g + h\| \leq \varepsilon$. But

$$(g + h)(y) = 1 + h(y) > 1,$$

and so $\|g + h\| > 1 > \varepsilon$. Therefore $\|g - h\| \leq \varepsilon$.

We set out to prove $\|g - h\| < \varepsilon$ and could have achieved this by replacing ε by δ with $0 < \delta < \varepsilon$ and $|1 - h(z)| < \delta^2/4$.

17. SOME NUMERICAL RANGE APPLICATIONS OF THE BISHOP-PHELPS-BOLLOBÁS THEOREM

As usual X will denote a Banach space.

Definition 1. Given $T \in B(X')$, the <u>lower numerical range</u> $LV(T)$ is defined by

$$LV(T) = \{(Tf)(x) : (x, f) \in \Pi(X)\}.$$

Theorem 2. <u>Let</u> $T \in B(X')$. <u>Then</u>

$$LV(T) \subset V(T) \subset LV(T)^-.$$

Proof. That $LV(T) \subset V(T)$ is obvious, for with $\hat{x}$ denoting the canonical image in X'' of $x \in X$, we have

$$(x, f) \in \Pi(X) \Rightarrow (f, \hat{x}) \in \Pi(X'),$$

and so $(Tf)(x) = \hat{x}(Tf) \in V(T)$.

Let $\lambda \in V(T)$, i.e. $\lambda = \Phi(Tf)$ with $(f, \Phi) \in \Pi(X')$. Let $0 < \varepsilon < 1$, and let X_1 denote the closed unit ball of X. Since $\hat{X}_1$ is weak* dense in X''_1, there exists $x \in X_1$ with

$$|\Phi(f) - \hat{x}(f)| < (\tfrac{\varepsilon}{2})^2, \qquad |\Phi(Tf) - \hat{x}(Tf)| < \varepsilon.$$

We have $x \in X_1$, $f \in S(X')$, and

$$|1 - f(x)| < (\tfrac{\varepsilon}{2})^2.$$

Therefore by the Bishop-Phelps-Bollobás theorem (Theorem 16.1) there exists $(y, g) \in \Pi(X)$ such that

$$\|y - x\| < \varepsilon, \qquad \|f - g\| < \varepsilon.$$

Then $(Tg)(y) \in LV(T)$ and

$$\begin{aligned} |\lambda-(Tg)(y)| &\leq |\lambda-\hat{x}(Tf)| + |(Tf)(x)-(Tf)(y)| + |(Tf)(y)-(Tg)(y)| \\ &< \varepsilon + \|T\|.\|x - y\| + \|T\| \|f - g\| \\ &\leq \varepsilon(1 + 2\|T\|). \end{aligned}$$

Since ε is arbitrary, this gives $\lambda \in LV(T)^-$.

Corollary 3. (Bollobás [115].) Let $T \in B(X)$. Then

$V(T) \subset V(T^*) \subset V(T)^-$.

Proof. $LV(T^*) = V(T)$.

The following example due to M. J. Crabb [130] shows that we can have $V(T) \neq V(T^*)$. It is not hard to produce a further example such that each inclusion in Corollary 3 is strict.

Example 4. Let $X = c_o$ and define $T \in B(X)$ by

$$(Tx)(n) = \sum_{k=0}^{\infty} 2^{-k-1}x(n + k) \qquad (n = 1, 2, 3, \ldots).$$

Then $V(T) \neq V(T^*)$.

Proof. It is clear that $\|Tx\| < 1$ for $\|x\| = 1$, and so $1 \notin V(T)$. With the normal identifications, $X' = l_1$, $X'' = l_\infty$, and

$$(T^*f)(n) = \sum_{k=1}^{n} 2^{k-n-1}f(k) \qquad (n = 1, 2, 3, \ldots).$$

Let $f \in X'$, $\phi \in X''$ be defined by

$$f(n) = \delta_n^1, \quad \phi(n) = 1 \qquad (n = 1, 2, 3, \ldots).$$

Then $(f, \phi) \in \Pi(X')$ and $1 = \phi(T^*f) \in V(T^*)$.

Theorem 5. Let $T \in B(X)$ with X a Banach space, let Y be a closed linear subspace of X that is invariant for T, let π denote the canonical mapping of X onto $Z = X - Y$, and let T_π be defined on Z by $T_\pi z = \pi Tx$ $(x \in z \in Z)$. Then

$$V(T_\pi) \subset V(T)^-.$$

Proof. Let $\lambda \in V(T_\pi)$. Then there exists $(z, f) \in \Pi(Z)$ with $\lambda = f(T_\pi z)$. Let $g = f \circ \pi$. Then $g \in X'$, $\|g\| \le 1$, and

$$g(x) = f(\pi x) = f(z) = 1 \qquad (x \in z).$$

Since $\inf\{\|x\| : x \in z\} = \|z\| = 1$, there exist $x_n \in z$ for which $\lim_{n\to\infty} \|x_n\| = 1$ and $g(x_n) = 1$. Thus $\|g\| = 1$, and by applying the Bishop-Phelps-Bollobás theorem to the pair $(\|x_n\|^{-1}x_n, g)$, we obtain $(y_n, h_n) \in \Pi(X)$ such that

$$\lim_{n\to\infty} \|y_n - x_n\| = \lim_{n\to\infty} \|h_n - g\| = 0.$$

Therefore

$$\lim_{n\to\infty} |h_n(Ty_n) - f(\pi Tx_n)| = \lim_{n\to\infty} |h_n(Ty_n) - g(Tx_n)| = 0.$$

Since $\pi Tx_n = T_\pi z$, this gives $f(T_\pi z) = \lim_{n\to\infty} h_n(Ty_n) \in V(T)^-$.

18. THE DEPENDENCE OF $\Pi(X)$ ON THE NORM

Let X with norm $\|\cdot\|$ be a non-zero Banach space, and let $N(X)$ denote the set of norms on X equivalent to the given norm $\|\cdot\|$. Corresponding to each choice of $p \in N(X)$ we obtain a subset $\Pi(X)$ of $X \times X'$ determined by the norm p, Π_p say. We show that with natural metrics on the space $N(X)$ and on the space of non-void bounded closed subsets of $X \times X'$ the mapping $p \to \Pi_p$ is continuous. The proof uses the Bishop-Phelps-Bollobás theorem.

We recall that norms p, q on X are said to be equivalent if there exist positive constants α, β such that

$$p(x) \le \alpha q(x), \quad q(x) \le \beta p(x) \qquad (x \in X).$$

Notation. Given $p, q \in N(X)$ and $\delta > 0$, we write

$$\frac{1}{\delta} \le \frac{p}{q} \le \delta$$

to denote that

$$\frac{1}{\delta} \le \frac{p(x)}{q(x)} \le \delta \quad (x \in X \setminus \{0\}),$$

and define $\mu(p, q)$, $d(p, q)$ by

$$\mu(p, q) = \inf\{\delta \ge 1 : \frac{1}{\delta} \le \frac{p}{q} \le \delta\},$$
$$d(p, q) = \log \mu(p, q).$$

Proposition 1. d is a metric on $N(X)$.

Proof. It is clear that for all $p, q, r \in N(X)$, $\mu(p, q) \ge 1$, $\mu(p, q) = \mu(q, p)$, $\mu(p, r) \le \mu(p, q)\,\mu(q, r)$; and it is easy to check that $\mu(p, q) = 1$ if and only if $p = q$.

Remark. $N(X)$ is complete with respect to the metric d, but we shall not use this.

Lemma 2. Given $p \in N(X)$, let p' denote the dual norm $p'(f) = \sup\{|f(x)| : p(x) \le 1\}$, $(f \in X')$. Then $p \to p'$ is an isometric mapping of $N(X)$ into $N(X')$.

Proof. If $p \le \delta q$, we have

$$|f(x)| \le p'(f)p(x) \le \delta p'(f)q(x) \quad (x \in X, \ f \in X'),$$

and so $q' \le \delta p'$.

Given $x \in X \setminus \{0\}$, there exists $f \in X'$ with $p'(f) = 1$ and $f(x) = p(x)$. Therefore, if $q' \le \delta p'$, we have

$$p(x) = f(x) \le q'(f)q(x) \le \delta q(x),$$

and so $p \le \delta q$.

This proves that $p \le \delta q$ if and only if $q' \le \delta p'$, and therefore $d(p', q') = d(p, q)$.

Notation. From this point on it will be convenient to denote also by p the dual norm of p. The set Π_p is then defined by

$$\Pi_p = \{(x, f) \in X \times X' : p(x) = p(f) = f(x) = 1\},$$

and given $T \in B(X)$, $V_p(T) = \{f(Tx) : (x, f) \in \Pi_p\}$.

Let M(E) denote the set of all non-void bounded closed subsets of a metric space E. Given $A, B \in M(E)$, define $d(A, B)$ by

$$d(A, B) = \max\{\sup_{x \in A} d(x, B), \sup_{x \in B} d(x, A)\}.$$

It is well known and easy to verify that d so defined is a metric on M(E). It is called the Hausdorff metric.

We use the given norm $\|\cdot\|$ on X to define a norm on $X \times X'$ by taking

$$\|(x, f)\| = \|x\| + \|f\| \qquad (x \in X, \ f \in X');$$

and then use the corresponding metric on $X \times X'$ to define the Hausdorff metric d on $M(X \times X')$.

We note that Π_p is a norm closed subset of $X \times X'$. For if $\lim_{n\to\infty} \|(x, f) - (x_n, f_n)\| = 0$ with $(x_n, f_n) \in \Pi_p$ $(n = 1, 2, \ldots)$, then $p(x) = \lim_{n\to\infty} p(x_n) = 1$, $p(f) = \lim_{n\to\infty} p(f_n) = 1$, and $f(x) = \lim_{n\to\infty} f_n(x_n) = 1$; $(x, f) \in \Pi_p$.

Theorem 3. With respect to the metrics defined on N(X), $M(X \times X')$ as above, the mapping $p \to \Pi_p$ is a continuous mapping of N(X) into $M(X \times X')$, and is uniformly continuous on each bounded subset of N(X).

Proof. Given $\kappa > 1$, let $G_\kappa = \{p \in N(X) : \mu(p, \|\cdot\|) < \kappa\}$; i.e. G_κ is the open ball in N(X) with centre $\|\cdot\|$ and radius $\log \kappa$. Given ε with $0 < \varepsilon < 1$ let

$$1 < \delta < 1 + (\tfrac{\varepsilon}{4})^2.$$

We prove that

$$p, q \in G_\kappa, \quad \mu(p,q) \le \delta \Rightarrow d(\Pi_p, \Pi_q) \le 3\kappa\varepsilon. \tag{1}$$

Since G_κ is open, this will prove that $p \to \Pi_p$ is a uniformly continuous mapping of G_κ into $M(\mathbf{X} \times \mathbf{X}')$; and, since $N(\mathbf{X}) = \cup\{G_\kappa : \kappa = 1, 2, \ldots\}$, this will complete the proof of the theorem.

Suppose then that $p, q \in G_\kappa$ and that $\mu(p,q) \le \delta$, so that

$$\frac{1}{\delta} \le \frac{p}{q} \le \delta. \tag{2}$$

Let $(x, f) \in \Pi_q$. Then, since $q(x) = q(f) = 1$, (2) gives

$$\frac{1}{\delta} \le p(x) \le \delta, \quad \frac{1}{\delta} \le p(f) \le \delta.$$

Let $z = (p(x))^{-1}x$, $h = (p(f))^{-1}f$. Then $p(z) = p(h) = 1$, and $h(z) = (p(x)p(f))^{-1}$. Therefore

$$\frac{1}{\delta^2} \le h(z) \le p(h)p(z) = 1. \tag{3}$$

Since $\delta < 2$, $\delta^2 - 1 < 3(\delta - 1) < 3(\frac{\varepsilon}{4})^2 < \varepsilon^2/4$. Therefore (3) gives

$$0 \le 1 - h(z) \le \frac{\delta^2 - 1}{\delta^2} < \varepsilon^2/4.$$

By the Bishop-Phelps-Bollobás theorem (Theorem 16.1), there exists $(y, g) \in \Pi_p$ with

$$p(y-z) < \varepsilon, \quad p(g-h) < \varepsilon. \tag{4}$$

Since $p(x) \le \delta$, we have $p(x) - 1 \le \delta - 1$; and, since $\frac{1}{\delta} \le p(x)$, we have $1 - p(x) \le 1 - \frac{1}{\delta} \le \delta - 1$. Therefore

$$|1 - p(x)| \le \delta - 1;$$

and so (4) gives

$$p(y - x) \le p(y - z) + p(z - x) < \varepsilon + p((\frac{1}{p(x)} - 1)x)$$
$$= \varepsilon + |1 - p(x)| \le \varepsilon + \delta - 1 < \tfrac{3}{2}\varepsilon.$$

Since $p \in G_\kappa$, we have $\|\cdot\| \le \kappa p$, and so

$$\|y - x\| < \tfrac{3}{2}\kappa\varepsilon.$$

Similarly $\|g - f\| < \frac{3}{2}\kappa\varepsilon$. Therefore

$$d((x, f), \Pi_p) < 3\kappa\varepsilon.$$

Since this holds for all $(x, f) \in \Pi_q$, we have

$$\sup\{d((x, f), \Pi_p) : (x, f) \in \Pi_q\} \leq 3\kappa\varepsilon.$$

A similar inequality holds with Π_p, Π_q interchanged, and so

$$d(\Pi_p, \Pi_q) \leq 3\kappa\varepsilon.$$

This proves (1), and completes the proof of the theorem.

Remark. The uniform continuity on G_κ proved in (1) above involves an explicit relation between ε and δ which is independent of the space **X**. It may be expressed in the form:

$$0 < \varepsilon < 3\kappa, \quad p, q \in G_\kappa, \quad d(p, q) < \frac{\varepsilon^2}{153\kappa^2} \Rightarrow d(\Pi_p, \Pi_q) < \varepsilon,$$

(for $\log(1 + (\varepsilon/4)^2) > \varepsilon^2/17 \quad (0 < \varepsilon < 1)$).

Corollary 4. Let $T \in B(X)$. With respect to the above defined metric on $N(X)$ and the Hausdorff metric on $M(\underset{\sim}{C})$, the mapping $p \to V_p(T)^-$ is continuous (and uniformly continuous on bounded subsets).

Proof. With $\kappa > 1$, define G_κ as in the proof of Theorem 3. Then $p \to \Pi_p$ is uniformly continuous on G_κ. Given $\varepsilon > 0$, choose $\delta > 0$ such that

$$p, q \in G_\kappa, \quad d(p, q) < \delta \Rightarrow d(\Pi_p, \Pi_q) < \varepsilon.$$

Given $\lambda \in V_q(T)$, we have $\lambda = f(Tx)$ for some $(x, f) \in \Pi_q$. Then $d((x, f), \Pi_p) < \varepsilon$ and so there exists $(y, g) \in \Pi_p$ with $\|x-y\| + \|f-g\| < \varepsilon$. Then

$$\begin{aligned} |g(Ty) - f(Tx)| &\leq |g(Ty) - f(Ty)| + |f(Ty) - f(Tx)| \\ &\leq \|g - f\| \, \|T\| \, \|y\| + \|f\| \, \|T\| \, \|y - x\| \\ &\leq \varepsilon \|T\| \kappa. \end{aligned}$$

Thus $\sup\{d(\lambda, V_p(T)^-) : \lambda \in V_q(T)\} \leq \varepsilon \|T\| \kappa$; and, by symmetry,

$$d(V_q(T)^-, V_p(T)^-) \leq \varepsilon \|T\| \kappa.$$

Remark. It is hoped that the results or methods of this section will lead to progress in the study of topological properties of $\Pi(X)$ and $V(T)$, for example through approximation to the given norm by smooth norms.

19. A THEOREM OF ZENGER

The stop press of NRI (§14, Remark (13)) drew attention to Chr. Zenger [78] in which it was proved that the convex hull of the point spectrum is contained in the spatial numerical range. The present section contains the proof of this remarkable result and of its extension due to M. J. Crabb [130].

Lemma 1. (Zenger.) Let $\|\cdot\|$ be a norm on $\underset{\sim}{C}^n$, let $\alpha_k > 0$ $(k = 1, \ldots, n)$, and let $\sum_{k=1}^{n} \alpha_k = 1$. Then there exists $\underset{\sim}{w} = (w_1, w_2, \ldots, w_n) \in \underset{\sim}{C}^n$ with $\|\underset{\sim}{w}\| = 1$ and $w_1 w_2 \ldots w_n \neq 0$ such that the functional ϕ on $\underset{\sim}{C}^n$ defined by

$$\phi(\underset{\sim}{z}) = \sum_{k=1}^{n} \alpha_k w_k^{-1} z_k \qquad (\underset{\sim}{z} = (z_1, \ldots, z_n) \in \underset{\sim}{C}^n)$$

satisfies

$$\|\phi\| \leq 1.$$

Remark. With $\underset{\sim}{w}$, ϕ as in the lemma, we obviously have $\phi(\underset{\sim}{w}) = 1$, and so

$$(\underset{\sim}{w}, \phi) \in \Pi(\underset{\sim}{C}^n).$$

Proof of Lemma 1. Consider the function F defined on $\underset{\sim}{C}^n$ by

$$F(\underset{\sim}{z}) = \prod_{k=1}^{n} |z_k|^{\alpha_k} \qquad (\underset{\sim}{z} = (z_1, \ldots, z_n) \in \underset{\sim}{C}^n),$$

and let

$$c = \sup\{F(\underset{\sim}{z}) : \|\underset{\sim}{z}\| \leq 1\}.$$

Clearly this supremum is attained at a point $\underset{\sim}{w}$ with $\|\underset{\sim}{w}\| \leq 1$, and since

$$F(\lambda \underset{\sim}{z}) = \lambda F(\underset{\sim}{z}) \qquad (\lambda \geq 0, \ \underset{\sim}{z} \in \underset{\sim}{C}^n),$$

we have $\|\underset{\sim}{w}\| = 1$. Since $\alpha_k > 0$, it is clear that $w_k > 0$ $(k=1, \ldots, n)$, for otherwise $F(\underset{\sim}{w}) = 0$.

Let $w_k = u_k + iv_k$ with $u_k, v_k \in \underset{\sim}{R}$, and regard F as a function of $2n$ real variables x_k, y_k,

$$F(\underset{\sim}{z}) = \prod_{k=1}^{n} (x_k^2 + y_k^2)^{\frac{1}{2}\alpha_k} \qquad (\underset{\sim}{z} = (x_1+iy_1, x_2+iy_2, \ldots, x_n+iy_n)).$$

Then F is differentiable at $\underset{\sim}{w}$, and

$$\frac{\partial F}{\partial x_k}\Big|_{\underset{\sim}{w}} = c\,\alpha_k u_k |w_k|^{-2}, \qquad \frac{\partial F}{\partial y_k}\Big|_{\underset{\sim}{w}} = c\,\alpha_k v_k |w_k|^{-2}.$$

Define a real linear functional ψ on $\underset{\sim}{C}^n$ by taking

$$\psi(\underset{\sim}{z}) = \sum_{k=1}^{n} \alpha_k |w_k|^{-2}(u_k x_k + v_k y_k) \qquad (\underset{\sim}{z} = (x_1+iy_1, \ldots, x_n+iy_n)).$$

Clearly we have $\psi(\underset{\sim}{w}) = 1$, and

$$F(\underset{\sim}{z}) - F(\underset{\sim}{w}) = c\psi(\underset{\sim}{z} - \underset{\sim}{w}) + o(\|\underset{\sim}{z} - \underset{\sim}{w}\|)$$

as $\underset{\sim}{z} \to \underset{\sim}{w}$. Thus

$$F(\underset{\sim}{z}) = c\psi(\underset{\sim}{z}) + o(\|\underset{\sim}{z} - \underset{\sim}{w}\|).$$

Suppose that there exists $\underset{\sim}{a} \in \underset{\sim}{C}^n$ such that $\|\underset{\sim}{a}\| \leq 1$ and $\psi(\underset{\sim}{a}) = 1 + \delta$ with $\delta > 0$. Then, with $0 < \alpha \leq 1$, we have

$$F(\alpha\underset{\sim}{a} + (1 - \alpha)\underset{\sim}{w}) = c\alpha(1 + \delta) + (1 - \alpha)c + o(\alpha) > c$$

for sufficiently small α. This contradicts the definition of c, and proves that $\|\psi\| \leq 1$.

Now consider the functional ϕ defined by

$$\phi(\underset{\sim}{z}) = \sum_{k=1}^{n} \alpha_k w_k^{-1} z_k .$$

We have

$$\begin{aligned}\phi(\underset{\sim}{z}) &= \sum_{k=1}^{n} \alpha_k |w_k|^{-2} w_k^* z_k \\ &= \sum_{k=1}^{n} \alpha_k |w_k|^{-2}(u_k x_k + v_k y_k) - i \sum_{k=1}^{n} \alpha_k |w_k|^{-2}(v_k x_k - u_k y_k) \\ &= \psi(\underset{\sim}{z}) - i\psi(i\underset{\sim}{z}).\end{aligned}$$

Then $\operatorname{Re} \phi = \psi$, and so, by Lemma 15.3, $\|\phi\| = \|\psi\| \le 1$.

Theorem 2. (Zenger [78].) Let X be a normed linear space over $\underset{\sim}{C}$, let $a_1, \ldots, a_r$ be linearly independent elements of X, and let $\alpha_k \ge 0$ $(k = 1, \ldots, r)$, $\sum_{k=1}^{r} \alpha_k = 1$. Then there exist $(x, f) \in \Pi(X)$, and complex numbers $w_1, \ldots, w_r$, such that $x = w_1 a_1 + \ldots + w_r a_r$ and

$$f(w_k a_k) = \alpha_k \qquad (k = 1, \ldots, r).$$

Proof. We suppose without loss of generality that we have a positive integer n with $1 \le n \le r$ such that $\alpha_k > 0$ $(1 \le k \le n)$, $\alpha_k = 0$ $(k > n)$. Define $P : \underset{\sim}{C}^n \to X$ by

$$P\underset{\sim}{z} = z_1 a_1 + \ldots + z_n a_n \qquad (\underset{\sim}{z} = (z_1, \ldots, z_n) \in \underset{\sim}{C}^n),$$

and define a norm on $\underset{\sim}{C}^n$ by taking

$$\|\underset{\sim}{z}\| = \|P\underset{\sim}{z}\| \qquad (\underset{\sim}{z} \in \underset{\sim}{C}^n).$$

Let $Y = P\underset{\sim}{C}^n$, and note that P maps $\underset{\sim}{C}^n$ isometrically and linearly onto Y. Let P^{-1} be the inverse mapping: $Y \to \underset{\sim}{C}^n$, and define f_o on Y by taking

$$f_o(y) = \phi(P^{-1}y) \qquad (y \in Y),$$

where $\underset{\sim}{w}$ and ϕ are as in Lemma 1. We have $P\underset{\sim}{w} \in Y$, $\|P\underset{\sim}{w}\| = 1$,

$$f_o(P\underset{\sim}{w}) = \phi(\underset{\sim}{w}) = 1,$$

and

$$|f_o(y)| = |\phi(P^{-1}y)| \leq \|P^{-1}y\| = \|y\| \qquad (y \in Y).$$

Thus $\|f_o\| = 1$. By the Hahn-Banach theorem, there exists $f \in X'$ with $f|_Y = f_o$ and $\|f\| = 1$. Take $x = P\underset{\sim}{w}$, i. e. $x = w_1 a_1 + \ldots + w_n a_n$. Then $\|x\| = \|\underset{\sim}{w}\| = 1$, and $f(x) = f_o(x) = f_o(P\underset{\sim}{w}) = 1$. Thus $(x, f) \in \Pi(X)$.

Let $\underset{\sim}{e}_k$ be the vector in $\underset{\sim}{C}^n$ with 1 in the k-th place and 0 in all other places. By definition of ϕ, we have

$$\phi(\underset{\sim}{e}_k) = \alpha_k w_k^{-1},$$

and, since $P\underset{\sim}{e}_k = a_k$, this gives

$$f_o(a_k) = \phi(P^{-1}a_k) = \phi(e_k) = \alpha_k w_k^{-1}.$$

Therefore $f(w_k a_k) = \alpha_k$ $(k = 1, 2, \ldots, n)$. Finally, we take $w_k = 0$ $(n < k \leq r)$.

Notation. Given a normed linear space X over $\underset{\sim}{C}$, and a linear mapping T of X into X we denote by p Sp(T) the <u>point spectrum</u> of T, i. e. the set of $\lambda \in \underset{\sim}{C}$ such that $Tx = \lambda x$ for some non-zero $x \in X$.

Theorem 3. (Zenger [78].) <u>Let</u> X <u>be a normed linear space over</u> $\underset{\sim}{C}$ <u>and let</u> T <u>be a linear mapping of</u> X <u>into</u> X. <u>Then</u> co p Sp(T) $\subset$ V(T).

Proof. Let $\lambda \in$ co p Sp(T). Then there exist distinct eigenvalues $\lambda_1, \ldots, \lambda_r \in$ p Sp(T) and $\alpha_1, \ldots, \alpha_r$, $\alpha_k \geq 0$ $(k = 1, \ldots, r)$, $\sum_{k=1}^{r} \alpha_k = 1$, such that $\lambda = \sum_{k=1}^{r} \alpha_k \lambda_k$. Let $a_1, \ldots, a_r$ be eigenvectors corresponding to $\lambda_1, \ldots, \lambda_r$. Then $a_1, \ldots, a_r$ are linearly independent, and, by Theorem 2, there exist $(x, f) \in \Pi(X)$ and complex numbers $w_1, \ldots, w_r$ such that $x = w_1 a_1 + \ldots + w_r a_r$ and $f(w_k a_k) = \alpha_k$ $(k = 1, \ldots, r)$. Then

$$f(Tx) = f(T(\sum_{k=1}^{r} w_k a_k)) = f(\sum_{k=1}^{r} w_k \lambda_k a_k) = \sum_{k=1}^{r} \alpha_k \lambda_k = \lambda.$$

Thus $\lambda \in V(T)$.

Theorem 4. (Crabb [130].) Let X be a Banach space over $\underset{\sim}{C}$, and let $T \in B(X)$. Then

$$\text{co } Sp(T) \subset V(T)^{-} .$$

Proof. We assume that $\|T\| \leq 1$. A moment's consideration of half-planes containing $\partial\, Sp(T)$ shows that $\text{co } Sp(T) = \text{co } \partial\, Sp(T)$. Since $\underset{\sim}{C}$ has dimension 2 over $\underset{\sim}{R}$, it follows by a theorem of Carathéodory that each point of $\text{co } Sp(T)$ is a convex combination of at most three points of $\partial\, Sp(T)$. We consider a convex combination of three distinct points of $\partial\, Sp(T)$, the case of two distinct points being done by a slightly simplified version of the proof, which we omit, and the case of one point being just the elementary fact that $\partial\, Sp(T) \subset V(T)^{-}$.

Suppose then that $\lambda = \alpha_1\lambda_1 + \alpha_2\lambda_2 + \alpha_3\lambda_3$ with $\lambda_1, \lambda_2, \lambda_3$ distinct points of $\partial\, Sp(T)$, $\alpha_k > 0$ $(k = 1, 2, 3)$, $\alpha_1 + \alpha_2 + \alpha_3 = 1$. Then

$$\Delta = \begin{vmatrix} 1 & 1 & 1 \\ \lambda_1 & \lambda_2 & \lambda_3 \\ \lambda_1^2 & \lambda_2^2 & \lambda_3^2 \end{vmatrix} \neq 0.$$

Let Δ_{ij} be the cofactor of the (i, j)th element of this determinant, let

$$\mu = \max\{|\Delta_{ij}|\,\Delta^{-1} : i, j = 1, 2, 3\},$$

and let $0 < \varepsilon < (9\mu)^{-1}$.

Since $\lambda_k \in \partial\, Sp(T)$, it belongs to the approximate point spectrum of T, and so there exists $x_k \in S(X)$ with

$$\|\lambda_k x_k - Tx_k\| < \varepsilon. \tag{1}$$

Since $|\lambda_k| \leq \|T\| \leq 1$, we have

$$\|\lambda_k^2 x_k - T^2 x_k\| \le \|\lambda_k(\lambda_k x_k - Tx_k)\| + \|T(\lambda_k x_k - Tx_k)\| \le 2\varepsilon. \quad (2)$$

By Theorem 2, there exists $(x, f) \in \Pi(X)$ with $x = w_1x_1 + w_2x_2 + w_3x_3$ and

$$f(w_k x_k) = \alpha_k \qquad (k = 1,\ 2,\ 3). \qquad (3)$$

Let $M = \max(|w_1|, |w_2|, |w_3|)$. Then, by (3) and (1),

$$\begin{aligned}|f(Tx) - \lambda| &= \left|\sum_{k=1}^{3} \{f(w_k Tx_k) - f(w_k x_k)\lambda_k\}\right| \\ &\le M \sum_{k=1}^{3} \|Tx_k - \lambda_k x_k\| \le 3\varepsilon M. \qquad (4)\end{aligned}$$

We prove that

$$M \le 3\mu(1 - 9\mu\varepsilon)^{-1}. \qquad (5)$$

Since μ depends only on $\lambda_1, \lambda_2, \lambda_3$ and ε is arbitrarily small, (4) and (5) will then show that $\lambda \in V(T)^-$ and complete the proof.

Let U denote the closed unit ball in X. Then, in turn,

$$\sum_{k=1}^{3} w_k x_k = x \in U, \qquad \sum_{k=1}^{3} w_k Tx_k \in U, \qquad \sum_{k=1}^{3} w_k T^2 x_k \in U.$$

Using (1) and (2), we therefore have

$$\begin{aligned}& w_1x_1 + w_2x_2 + w_3x_3 \in U, \\ & w_1\lambda_1x_1 + w_2\lambda_2x_2 + w_3\lambda_3x_3 \in (1 + 3M\varepsilon)U, \\ & w_1\lambda_1^2x_1 + w_2\lambda_2^2x_2 + w_3\lambda_3^2x_3 \in (1 + 6M\varepsilon)U.\end{aligned}$$

Therefore

$$\begin{aligned}& \Delta w_1 x_1 \in \Delta_{11}U + \Delta_{21}(1 + 3M\varepsilon)U + \Delta_{31}(1 + 6M\varepsilon), \\ & |w_1| \le \mu(3 + 9M\varepsilon).\end{aligned}$$

Thus $M \le 3\mu + 9\mu\varepsilon M$, and (5) is proved.

Corollary 5. Let $(X, \|\cdot\|)$ be a Banach space over $\underset{\sim}{C}$ and let $N(X)$ denote the set of all norms on X equivalent to $\|\cdot\|$. Given $p \in N(X)$, let $V_p(T)$ denote the spatial numerical range of T corresponding to the norm p. Then

$$\text{co } Sp(T) = \cap \{V_p(T)^- : p \in N(X)\}.$$

Proof. By NRI Theorem 10.4 and Theorem 4, we have

$$\text{co } Sp(T) \subset \cap\{V_p(T)^- : p \in N(X)\} \subset \cap\{\overline{\text{co}}\ V_p(T) : p \in N(X)\} = \text{co } Sp(T).$$

Corollary 6. Let X be a Banach space over $\underset{\sim}{C}$ and let $T \in B(X')$. Then

$$\text{co } Sp(T) \subset LV(T)^-.$$

Proof. By Theorem 4 and Theorem 17.2 we have

$$\text{co } Sp(T) \subset V(T)^- = LV(T)^-.$$

Remark. We need not have $p\ Sp(T) \subset LV(T)$; for example let $X = c_o$ and let $T \in B(X')$ be defined by

$$(Tf)(n) = 2^{-n} \sum_{k=1}^{\infty} f(k) \qquad (f \in l_1).$$

It is readily verified that $1 \in p\ Sp(T) \backslash LV(T)$.

20. EIGENVALUES IN THE BOUNDARY OF THE NUMERICAL RANGE

Throughout this section, let X be a normed linear space over $\underset{\sim}{C}$, let $T \in B(X)$, and let $Ker(T) = \{x \in X : Tx = 0\}$.

The theorem of Nirschl and Schneider (NRI Theorem 10.10) states that each eigenvalue in the boundary of $V(B(X), T)$ has ascent 1. There is no real loss of generality in taking the eigenvalue to be 0, and the theorem then takes the form:

$$0 \in p\ Sp(T) \cap \partial V(B(X), T) \Rightarrow Ker(T^2) = Ker(T). \tag{1}$$

A stronger conclusion from the same hypothesis was obtained by Sinclair [202], namely that Ker(T) is orthogonal to TX in a sense meaningful for normed linear spaces which will now be defined.

Definition 1. Let A, B be linear subspaces of X. Then A is said to be orthogonal to B (in the sense of G. Birkhoff) and we write $A \perp B$ if

$$\|a + b\| \geq \|a\| \qquad (a \in A,\ b \in B).$$

In terms of this concept the theorem of Sinclair states:

$$0 \in p\,Sp(T) \cap \partial V(B(X), T) \Rightarrow Ker(T) \perp TX. \tag{2}$$

It is easy to see that (1) follows from (2); in fact we have the relations shown in the following lemma.

Lemma 2. The implications

$$(iv) \Rightarrow (iii) \Rightarrow (ii) \Longleftrightarrow (i)$$

hold between the statements:

(i) $Ker(T^2) = Ker(T)$,

(ii) $Ker(T) \cap TX = \{0\}$,

(iii) $Ker(T) \cap (TX)^- = \{0\}$,

(iv) $Ker(T) \perp TX$.

Proof. If (iv) holds and $x \in Ker(T) \cap (TX)^-$, then $x = \lim_{n\to\infty} y_n$ with $y_n \in TX$. Since $x \in Ker(T)$ and $-y_n \in TX$, we have $\|x-y_n\| \geq \|x\|$, $x = 0$. The rest is obvious.

So far these results involved the numerical range V(B(X), T), i. e. $\overline{co}\, V(T)$, and it was natural to ask whether similar but potentially more powerful results could be obtained for the spatial numerical range V(T). The crucial step was taken by Crabb [131] in proving that eigenvalues in $\partial V(T)$ have ascent 1. Very recently, Crabb and Sinclair [136] have proved that $Ker(T) \perp TX$ when $0 \in p\,Sp(T) \cap \partial V(T)$. These results for V(T) seem to lie deeper than the corresponding results for V(B(X), T), and the only proof known to us uses a generalization of the Brouwer fixed point theorem due to Kakutani [170].

The results for $V(T)$ contain those for $V(B(X), T)$, but, because it is useful to have available simple proofs of simple theorems and because the proof is a thing of beauty, we give a proof of (2) which is close to the proof originally given by Sinclair.

Lemma 3. The following statements are equivalent.

(i) $\operatorname{Ker}(T) \perp TX$.

(ii) $D(X,u) \cap \operatorname{Ker}(T^*) \neq \phi \quad (u \in S(X) \cap \operatorname{Ker}(T))$.

Proof. (i) $\Rightarrow$ (ii). Let $u \in S(X) \cap \operatorname{Ker}(T)$, and suppose that $\operatorname{Ker}(T) \perp TX$. Then

$$|\lambda| = \|\lambda u\| \leq \|\lambda u + Tx\| \qquad (\lambda \in \underset{\sim}{C}, \quad x \in X).$$

Define f_o on $\underset{\sim}{C}u + TX$ by taking $f_o(\lambda u + Tx) = \lambda$. Then $\|f_o\| \leq 1$, $f_o(u) = 1$, $f_o(TX) = \{0\}$. By the Hahn-Banach theorem, f_o can be extended to give an element of $D(X,u) \cap \operatorname{Ker}(T^*)$.

(ii) $\Rightarrow$ (i). Let $u \in S(X) \cap \operatorname{Ker}(T)$ and $f \in D(X,u) \cap \operatorname{Ker}(T^*)$. Then $\|u + Tx\| \geq |f(u + Tx)| = 1 = \|u\|$. Thus $\operatorname{Ker}(T) \perp TX$.

Theorem 4. Let $0 \in \partial G \cap p\,Sp(T)$, where G is some open circular disc with $G \cap V(T) = \phi$. Then

$\operatorname{Ker}(T) \perp TX$.

Proof. By multiplying T by a suitable non-zero scalar we may arrange that G is the disc $\{\zeta \in \underset{\sim}{C} : |\zeta - 1| < 1\}$. Then by NRI Theorem 10.1 and its proof, $I - T$ is invertible and

$$\|(I - T)x\| \geq \|x\| \qquad (x \in X).$$

Therefore $\|(I - T)^{-1}\| \leq 1$.

Let $u \in S(X) \cap \operatorname{Ker}(T)$. Then

$$((I - T)^{-1})^* D(X,u) \subset D(X,u).$$

For $(I - T)u = u$ gives $(I - T)^{-1}u = u$; and so, with $f \in D(X, u)$,

$$((I - T)^{-1})^*f(u) = f((I - T)^{-1}u) = f(u) = 1;$$

and also $\|((I - T)^{-1})^*f\| \leq 1$. The set $D(X, u)$ is convex, and is compact in the weak* topology, and $((I - T)^{-1})^*$ is weak* continuous and linear. Therefore, by either the Schauder-Tychonoff or the Markoff-Kakutani fixed point theorem (see Dunford and Schwartz [148], p. 456) there exists $g \in D(X, u)$ with $((I - T)^{-1})^*g = g$. But then $(I - T)^*g = g$, $T^*g = 0$. We have proved that statement (ii) of Lemma 3 holds, and therefore $\mathrm{Ker}(T) \perp TX$.

Corollary 5. $0 \in \mathrm{p\,Sp}(T) \cap \partial V(B(X), T) \Rightarrow \mathrm{Ker}(T) \perp TX$.

Proof. Let $0 \in \mathrm{p\,Sp}(T) \cap \partial V(B(X), T)$. Since $V(B(X), T) = \overline{\mathrm{co}}\, V(T)$, 0 is on the boundary of an open half plane which does not intersect $V(T)$.

In very recent work not yet published Crabb and Sinclair have found an entirely elementary proof of a strengthened form of Corollary 5, as follows.

Theorem 6. <u>If</u> 0 <u>is not in the interior of</u> $V(B(X), T)$, <u>then</u>

$$\|x + Ty\| \geq \|x\| - \sqrt{8\|Tx\| \, \|y\|} \qquad (x, y \in X). \tag{3}$$

Proof. By multiplying T by a suitable complex number of modulus 1, we may arrange that $V(B(X), T)$ is contained in the left hand half plane $\{\zeta \in \mathbb{C} : \mathrm{Re}\, \zeta \leq 0\}$, without disturbing the inequality (3). The inequality (3) is obvious if $x = 0$. Let $x, y \in X$ with $x \neq 0$.

Choose $\mu \in \mathbb{C}$ with $x + \mu y \neq 0$, and take $z = \|x + \mu y\|^{-1}(x + \mu y)$. Choose $f \in D(X, z)$. Then $\mathrm{Re}\, f(Tz) \leq 0$, and so for all $\alpha \geq 0$, we have

$$\alpha \leq |f(Tz) - \alpha| = |f((T - \alpha I)z)| \leq \|(T - \alpha I)z\|.$$

Thus

$$\alpha\|x + \mu y\| \leq \|(T - \alpha I)(x + \mu y)\| \qquad (\mu \in \mathbb{C}, \quad \alpha \geq 0).$$

Taking $\mu = -\alpha$ we now obtain, for all $\alpha \geq 0$,

$$\alpha \|x\| - \alpha^2 \|y\| \leq \|\alpha^2 y + Tx - \alpha(x + Ty)\|$$
$$\leq \alpha^2 \|y\| + \|Tx\| + \alpha \|x + Ty\|.$$

We have proved that

$$A\alpha^2 + B\alpha + C \geq 0 \qquad (\alpha \geq 0), \tag{4}$$

with $A = 2\|y\|$, $B = \|x + Ty\| - \|x\|$, $C = \|Tx\|$. If $B \geq 0$, the inequality (3) is obviously satisfied. If $B < 0$, we take $\alpha = -B/2A$ in (4) and obtain $4AC \geq B^2$, which is the required inequality (3).

A method of proof similar to the proof of Theorem 4, but using the Ryll-Nardzewski fixed point theorem yields the following theorem on Hermitian operators on a reflexive Banach space.

Theorem 7. Let X be a reflexive Banach space, and let $u \in S(X)$. Then there exists $f \in D(X, u)$ such that $T^*f = 0$ for all Hermitian operators $T \in B(X)$ for which $Tu = 0$.

Proof. $D(X, u)$ is a non-void weakly compact subset of X'. Let U denote the set of all linear isometries A of X onto X such that $Au = u$. Then

$$A^*D(X, u) \subset D(X, u) \qquad (A \in U),$$

and the Ryll-Nardzewski fixed point theorem [187] is applicable. Thus there exists $f \in D(X, u)$ with

$$A^*f = f \qquad (A \in U).$$

Let T be a Hermitian operator on X with $Tu = 0$. Then

$$\exp(i\alpha T) \in U \qquad (\alpha \in \mathbb{R}),$$

and so

$$(\exp(i\alpha T))^*f = f \qquad (\alpha \in \mathbb{R}).$$

This gives, for all $\alpha \in \mathbb{R}$,

$$0 = i\alpha T^*f + \frac{(i\alpha)^2}{2!} T^{*2}f + \dots ,$$

and so $T^*f = 0$.

Remarks. (1) Theorem 7 is not valid for a general Banach space X, as the following example, which we owe to A. M. Sinclair, shows. Take $X = B(H)$ with H a Hilbert space, let $A \in B(H)$ be self-adjoint, and let δ_A be the operator defined on X by

$$\delta_A T = (AT - TA) \qquad (T \in X).$$

Then δ_A is a Hermitian operator on X; for

$$\begin{aligned} \|(I + i\alpha\delta_A)T\| &= \|\tfrac{1}{2}(I + 2i\alpha A)T + \tfrac{1}{2}T(I - 2i\alpha A)\| \\ &\le (\tfrac{1}{2}\|I + 2i\alpha A\| + \tfrac{1}{2}\|I - 2i\alpha A\|)\|T\|, \end{aligned}$$

and so

$$1 \le \|I + i\alpha\delta_A\| \le 1 + o(\alpha) \qquad (\alpha \in \mathbb{R}),$$

since A is Hermitian. If there exists $f \in D(X, I)$ with $\delta_A^* f = 0$ for all self-adjoint operators $A \in B(H)$, we have

$$f(AT - TA) = 0 \qquad (A, T \in B(H)).$$

But this then gives $f(I) = 0$ (see Halmos [30], Problem 186, Corollary 2).

(2) Theorem 7 holds for some non-reflexive spaces, for we may replace the Ryll-Nardzewski theorem by the Markov-Kakutani theorem provided $H(B(X))$ is commutative. This last condition is satisfied if $X = C(E)$, E compact Hausdorff, and also for other classes of function spaces.

(3) Is Theorem 7 true with Hermitian T replaced by dissipative T? If T is dissipative and $Tu = 0$, we have still

$$(\exp(i\alpha T))^* D(X, u) \subset D(X, u) \qquad (\alpha \ge 0). \qquad (5)$$

However the 'non-contractive' condition in the Ryll-Nardzewski theorem is not satisfied, and so the proof fails. The result however remains an immediate consequence of (5) if $D(X, u)$ is a singleton, in particular if X is a Hilbert space.

For the proof of the main Crabb-Sinclair theorem we shall need the following generalized fixed point theorem for the special case in which the normed linear space is $\underset{\sim}{C}$.

Theorem 8. (Kakutani [170].) Let E be a compact convex set in a finite dimensional normed linear space, and let $K(E)$ denote the set of all non-void closed convex subsets of E. Let $\phi : E \to K(E)$ be an upper semi-continuous mapping. Then there exists $x_o \in E$ such that

$$x_o \in \phi(x_o).$$

Proof. We show first that the proof can be reduced to the special case in which E is a simplex in $\underset{\sim}{R}^r$. First, since all norms on a finite dimensional linear space are equivalent there is no loss of generality in supposing that the normed linear space is $\underset{\sim}{R}^r$ with the Euclidean norm, for which the unit ball is rotund. Let Δ be a simplex with $E \subset \Delta$. Since E is compact and convex, each $x \in \Delta$ has a well defined nearest point $\theta(x) \in E$; moreover the mapping θ is continuous. Then the mapping $\phi \circ \theta$ of Δ into $K(\Delta)$ satisfies the conditions of the theorem with E replaced by Δ. If the theorem is proved for Δ we then have $x_o \in \Delta$ with $x_o \in (\phi \circ \theta)(x_o)$. Since $(\phi \circ \theta)(x_o) \in K(E)$, this gives $x_o \in E$. But then $\theta(x_o) = x_o$, and so $x_o \in \phi(x_o)$.

Suppose then that E is a simplex in $\underset{\sim}{R}^r$. For each positive integer n, let $\sigma^{(n)}$ denote the set of vertices of the nth barycentric simplicial subdivision of the simplex E. For each $x \in \sigma^{(n)}$ choose a point $\phi_n(x)$ arbitrarily in $\phi(x)$, and extend the mapping ϕ_n by linearity within each subsimplex with its vertices in $\sigma^{(n)}$. In this way we obtain a continuous mapping $\phi_n : E \to E$.

By the Brouwer fixed point theorem, there exists $u_n \in E$ such that

$$\phi_n(u_n) = u_n.$$

Let E_n be an r-dimensional subsimplex in the nth subdivision such that $u_n \in E_n$, and let $x_o^n, \ldots, x_r^n$ be the vertices of E_n. We have

$$u_n = \alpha_o^n x_o^n + \ldots + \alpha_r^n x_r^n,$$

with $\alpha_j^n \geq 0$ and $\sum_{j=0}^{r} \alpha_j^n = 1$. Let

$$\phi_n(x_j^n) = y_j^n \quad (j = 0, \ldots, r).$$

By compactness, there exists a strictly increasing sequence $\{n_k\}$ of positive integers such that

$$\lim_{k\to\infty} u_{n_k} = u, \quad \lim_{k\to\infty} \alpha_j^{n_k} = \alpha_j, \quad \lim_{k\to\infty} y_j^{n_k} = y_j \quad (j = 0, \ldots, r).$$

Since $u_{n_k} \in E_{n_k}$ and the diameter of E_{n_k} tends to zero, we have

$$\lim_{k\to\infty} x_j^{n_k} = u.$$

We have $y_j^{n_k} \in \phi(x_j^{n_k})$, and therefore the closed graph criterion (Lemma 15. 7) gives

$$y_j \in \phi(u) \quad (j = 0, \ldots, r). \tag{6}$$

We have $\alpha_j \geq 0$, $\sum_{j=0}^{r} \alpha_j = 1$, and

$$u = \lim_{k\to\infty} u_{n_k} = \lim_{k\to\infty} \phi_{n_k}(u_{n_k}) = \lim_{k\to\infty} \alpha_o^{n_k} y_o^{n_k} + \ldots + \alpha_r^{n_k} y_r^{n_k} = \alpha_o y_o + \ldots + \alpha_r y_r.$$

Thus $u = \alpha_o y_o + \ldots + \alpha_r y_r \in \phi(u)$, by (6) and the convexity of $\phi(u)$.

Theorem 9. (Crabb-Sinclair.) Let $x, y \in X$, suppose that

$$\varepsilon = \|x\| - \|x + Ty\| > (8\|Tx\| \|y\|)^{\frac{1}{2}}, \tag{7}$$

and let

$$\gamma = \frac{\varepsilon^2 - 8\|Tx\|\ \|y\|}{2\|y\|(4\|x\| + \varepsilon)}\ . \tag{8}$$

Then V(T) contains the closed circular disc $\{\zeta \in \underset{\sim}{C} : |\zeta| \le \gamma\}$.

In brief, 0 is an interior of V(T) whenever the inequality (7) holds for some x, y ∈ X.

Proof. Take $\delta = \varepsilon(4\|y\|)^{-1}$, $\Delta = \{\zeta \in \underset{\sim}{C}, |\zeta| \le 1\}$, and let $\alpha \in \gamma\Delta$.

Given $\lambda \in \delta\Delta$, we have

$$\|x+\lambda y+Ty\| \le \|x\| - \varepsilon + \delta\|y\| \le \|x+\lambda y\| - \varepsilon + 2\delta\|y\| = \|x+\lambda y\| - \frac{\varepsilon}{2}, \tag{9}$$

from which $\|x + \lambda y\| > 0$. Let $w(\lambda) = \|x + \lambda y\|^{-1}(x + \lambda y)$, and

$$F(\lambda) = \lambda + V(T, w(\lambda)) - \alpha \qquad (\lambda \in \delta\Delta).$$

We prove that $F(\lambda) \subset \delta\Delta$. Each point $\mu \in V(T, w(\lambda))$ is of the form $\mu = f(Tw(\lambda))$ with $f \in D(w(\lambda))$. Thus $\mu\|x + \lambda y\| = f(Tx + \lambda Ty)$, and so the inequality (9) gives

$$\begin{aligned} \|x + \lambda y\|\ |\lambda + \mu| &= |\lambda f(x + \lambda y) + f(Tx + \lambda Ty)| \\ &= |\lambda f(x + \lambda y + Ty) + f(Tx)| \\ &\le \delta\|x + \lambda y + Ty\| + \|Tx\| \\ &\le \delta\|x + \lambda y\| - \frac{\varepsilon\delta}{2} + \|Tx\|. \end{aligned} \tag{10}$$

By (8), we have

$$\|Tx\| = \frac{\varepsilon^2}{8\|y\|} - \gamma(\|x\| + \frac{\varepsilon}{4}) = \frac{\varepsilon\delta}{2} - \gamma(\|x\| + \frac{\varepsilon}{4}),$$

and so

$$\gamma\|x + \lambda y\| + \|Tx\| - \frac{\varepsilon\delta}{2} \le \gamma\|x\| + \gamma\delta\|y\| - \gamma(\|x\| + \frac{\varepsilon}{4}) = 0.$$

With (10), this gives

$$|\lambda + \mu| \le \delta + \frac{\|Tx\| - \frac{\varepsilon\delta}{2}}{\|x + \lambda y\|} \le \delta - \gamma,$$

$$|\lambda + \mu - \alpha| \leq |\lambda + \mu| + |\alpha| \leq \delta - \gamma + \gamma = \delta.$$

We have now proved that F is a mapping of $\delta\Delta$ into the non-void compact convex subsets of $\delta\Delta$. By Lemma 15. 8 and the continuity of the mapping $\lambda \to w(\lambda)$, F is upper semi-continuous. Therefore, by Theorem 8, there exists $\lambda \in \delta\Delta$ with $\lambda \in F(\lambda)$, i. e.

$$0 \in V(T, w(\lambda)) - \alpha.$$

This proves that $\alpha \in V(T)$, and so $\gamma\Delta \subset V(T)$.

Corollary 10. $0 \in p\,Sp(T) \cap \partial V(T) \Rightarrow Ker(T) \perp TX.$

Proof. Take $x \in Ker(T)$ in Theorem 9.

Corollary 11. $0 \in p\,Sp(T) \cap \partial V(T) \Rightarrow Ker(T^2) = Ker(T).$

Proof. Lemma 2 and Corollary 10.

Corollary 12. _If_ 0 _is not an interior point of_ V(T), _then_

$$\|T\|^2 \leq 8\|T^2\|.$$

Proof. Suppose that 0 is not an interior point of V(T). Then, by Theorem 9,

$$\|x\| - \|x + Ty\| \leq (8\|Tx\| \|y\|)^{\frac{1}{2}} \qquad (x, y \in X).$$

Given $y \in X$, take $x = -Ty$, and we have

$$\|Ty\|^2 \leq 8\|T^2 y\| \|y\| \qquad (y \in X).$$

Corollary 13. _Suppose that_ $0 \notin \cup \{\text{int } V(T^{2^{n-1}}) : n = 1, 2, \ldots\}$. _Then_

$$\|T\| \leq 8\rho(T).$$

Proof. By Corollary 12 applied to $T^{2^{n-1}}$, we have

$$\|T^{2^{n-1}}\|^2 \leq 8\|T^{2^n}\|.$$

Remark. In particular, if T is quasinilpotent and T^{2^n} is dissipative for $n = 0, 1, 2, \ldots$, then $T = 0$. Recall that if S is the Volterra integration operator on $L^2[0,1]$ then $-S$ is quasinilpotent and dissipative (see Halmos [30, pp. 166, 167]); in this case $0 \in \operatorname{int} V(S^2)$.

Corollary 14. Let X have finite dimension, and let $\operatorname{int} V(T) = \phi$. Then there exist complex numbers α, β with $|\beta| = 1$ such that $\alpha I + \beta T$ can be represented by a real diagonal matrix. If also $\mathrm{Sp}(T) = \{\lambda\}$, a singleton, then $T = \lambda I$.

Proof. By Zenger's theorem (Theorem 19. 3), $\operatorname{co} \mathrm{Sp}(T) \subset V(T)$, $\operatorname{int} \operatorname{co} \mathrm{Sp}(T) = \phi$, and so $\mathrm{Sp}(T)$ is contained in a straight line segment. Thus there exist $\alpha, \beta \in \underset{\sim}{C}$ with $|\beta| = 1$ such that $\mathrm{Sp}(\alpha I + \beta T) \subset \underset{\sim}{R}$. Also each eigenvalue of T is in the boundary of V(T), and so by Corollary 11 has ascent 1. Therefore with respect to a suitably chosen basis, $\alpha I + \beta T$ is a real diagonal matrix. If also $\mathrm{Sp}(T) = \{\lambda\}$, the diagonal elements of this matrix are all $\alpha + \beta\lambda$.

Under the hypotheses of Corollary 14 one can show further that V(T) is a line segment when Sp(T) has two points; this establishes one further case of the general conjecture:

$$\operatorname{int} V(T) = \phi \Rightarrow V(T) \text{ is a line segment.}$$

Other special cases of the conjecture may be deduced from the result of M. J. Crabb below; for example, V(T) is a line segment if V(T) is a convex curve.

Theorem 15. (Crabb [131].) Let $T \in B(X)$ with

$$\operatorname{Im} \lambda \geq 0 \qquad (\lambda \in V(T)),$$

and let $p(T) = 0$ for some non-zero complex polynomial p with all its zeros real. Then T is Hermitian.

Proof. The eigenvalues of T are real and so lie on the boundary of V(T). By Corollary 11 (or NRI Theorem 10. 10), they have ascent one. We may thus suppose that p has simple zeros. Let

$$p(z) = c_0 + c_1 z + \ldots + c_k z^k$$

with distinct zeros $\alpha_1, \ldots, \alpha_k$. There exist $P_1, \ldots, P_k \in B(X)$ with

$$T^r = \alpha_1^r P_1 + \ldots + \alpha_k^r P_k \qquad (r = 0, 1, \ldots, k-1).$$

The recurrence relations for the powers of T then give

$$T^r = \alpha_1^r P_1 + \ldots + \alpha_k^r P_k \qquad (r = k, k+1, \ldots)$$

and so

$$e^{\lambda T} = e^{\lambda \alpha_1} P_1 + \ldots + e^{\lambda \alpha_k} P_k \qquad (\lambda \in \underset{\sim}{C}).$$

Given $\varepsilon > 0$, by Kronecker's theorem there exist integers $m, m_1, \ldots, m_k$ with $m > 0$ and

$$|m\alpha_r - m_r| < \varepsilon \qquad (r = 1, 2, \ldots, k).$$

Therefore, given $0 < t < 2\pi$,

$$|e^{i(2\pi m - t)\alpha_r} - e^{-it\alpha_r}| < 2\pi\varepsilon \qquad (r = 1, 2, \ldots, k),$$

and so $\|e^{i(2\pi m-t)T} - e^{-itT}\| < K\varepsilon$, where $K = 2\pi(\|P_1\| + \ldots + \|P_k\|)$. Since $\sup \operatorname{Re} V(iT) = 0$ we have $\|e^{ixT}\| \le 1$ $(x > 0)$ and hence $\|e^{-itT}\| \le 1 + K\varepsilon$. Since ε was arbitrary this gives $\|e^{-itT}\| \le 1$ and so $\sup \operatorname{Re} V(-iT) \le 0$. The proof is complete.

21. ABSOLUTE NORMS ON C^2

In order to establish geometrical and topological properties of spatial numerical ranges for operators on X we require an explicit description of $\Pi(X)$. Such a description is available for some of the classical Banach spaces. In this section we characterize absolute norms on $\underset{\sim}{C}^2$ in terms of convex functions, and for such norms we determine $\Pi(\underset{\sim}{C}^2)$ and $V(T)$ for an arbitrary operator on $\underset{\sim}{C}^2$. This characterization is the basis for the construction of several special examples of numerical ranges.

We recall that a linear norm $\|\cdot\|$ on $\underset{\sim}{C}^2$ is <u>absolute</u> if

$$\|(z, w)\| = \|(|z|, |w|)\| \qquad (z, w \in \mathbb{C})$$

and normalized if

$$\|(1, 0)\| = \|(0, 1)\| = 1.$$

We write N_a for the family of absolute normalized linear norms on $\mathbb{C}^2$. We denote the l_p norms $(1 \le p \le \infty)$ on $\mathbb{C}^2$ by $\|\cdot\|_p$. Clearly $\|\cdot\|_p \in N_a$.

Lemma 1. For $\|\cdot\| \in N_a$, $\|\cdot\|_\infty \le \|\cdot\| \le \|\cdot\|_1$.

Proof. Given $z, w \in \mathbb{C}$, we have

$$\begin{aligned}
\|(z, w)\|_\infty &= \max\{\|(z, 0)\|, \|(0, w)\|\} \\
&\le \tfrac{1}{2}\max\{\|(z, w)\| + \|(z, -w)\|, \|(z, w)\| + \|(-z, w)\|\} \\
&= \|(z, w)\| \\
&\le \|(z, 0)\| + \|(0, w)\| \\
&= \|(z, w)\|_1.
\end{aligned}$$

Lemma 2. Let $\|\cdot\| \in N_a$.

(i) $|p| \le |r|,\ |q| \le |s| \Rightarrow \|(p, q)\| \le \|(r, s)\|$.

(ii) $|p| < |r|,\ |q| < |s| \Rightarrow \|(p, q)\| < \|(r, s)\|$.

Proof. We may suppose $p, q, r, s \ge 0$.

(i) Since (p, q) belongs to the line segment $[(r, q), (-r, q)]$ we have

$$\|(p, q)\| \le \|(r, q)\|,$$

and similarly $\|(r, q)\| \le \|(r, s)\|$.

(ii) We may suppose (by symmetry) that $qr \le ps$. Then

$(p, q) \in [(p, \frac{s}{r}p), (p, -\frac{s}{r}p)]$ and so

$$\begin{aligned}
\|(p, q)\| &\le \|(p, \tfrac{s}{r}p)\| \\
&= \tfrac{p}{r}\|(r, s)\| \\
&< \|(r, s)\|.
\end{aligned}$$

An absolute norm $\|\cdot\|$ on $\underset{\sim}{C}^2$ is clearly determined by its values on the line segment joining $(1, 0)$ to $(0, 1)$. We define

$$\psi(t) = \|(1-t, t)\| \qquad (0 \le t \le 1). \qquad (1)$$

Then ψ is continuous and convex on $[0, 1]$, $\psi(0) = \psi(1) = 1$, and by Lemma 1, we have

$$\max\{1-t, t\} \le \psi(t) \le 1 \qquad (0 \le t \le 1).$$

Let Ψ denote the family of all functions on $[0, 1]$ with the above properties.

Lemma 3. N_a and Ψ are in 1-1 correspondence under equation (1).

Proof. Given $\psi \in \Psi$ define $\|\cdot\|$ on $\underset{\sim}{C}^2$ by $\|(0, 0)\| = 0$, and

$$\|(z, w)\| = (|z| + |w|)\, \psi\left(\frac{|w|}{|z| + |w|}\right) \qquad ((z, w) \ne (0, 0)).$$

Since $\psi \ge \frac{1}{2}$, all the properties of an absolute normalized norm are clear except the triangle inequality. For this it is enough to show that

$$\|(a+c, b+d)\| \le \|(a, b)\| + \|(c, d)\|$$

when each vector is non-zero. We show first that

$$(p + q)\, \psi\left(\frac{q}{p+q}\right) \le (r + s)\, \psi\left(\frac{s}{r+s}\right) \qquad (0 < p \le r,\ 0 < q \le s). \qquad (2)$$

Since ψ is continuous and convex with $\psi(1) = 1$, $\psi(t) \ge t$, the function $t \to \psi(t)/t$ is non-increasing and therefore

$$(p + q)\, \psi\left(\frac{q}{p+q}\right) \le (r + q)\, \psi\left(\frac{q}{r+q}\right).$$

A similar argument gives

$$(r + q)\, \psi\left(\frac{q}{r+q}\right) \le (r + s)\, \psi\left(\frac{s}{r+s}\right).$$

Finally, by (2) and the convexity of ψ,

$$\|(a+c,\ b+d)\| = (|a+c| + |b+d|)\,\psi(\frac{|b+d|}{|a+c| + |b+d|})$$

$$\leq (|a| + |b| + |c| + |d|)\,\psi(\frac{|b|+|d|}{|a|+|b|+|c|+|d|})$$

$$\leq (|a| + |b|)\,\psi(\frac{|b|}{|a|+|b|}) + (|c| + |d|)\,\psi(\frac{|d|}{|c|+|d|})$$

$$= \|(a,\ b)\| + \|(c,\ d)\|.$$

We recall that a continuous convex function ψ has left and right derivatives ψ'_L, ψ'_R. Let G be defined on $[0, 1]$ by

$$G(0) = [-1,\ \psi'_R(0)], \qquad G(1) = [\psi'_L(1),\ 1],$$

$$G(t) = [\psi'_L(t),\ \psi'_R(t)] \qquad (0 < t < 1).$$

Note that $G(t)$ is the set of gradients of tangents to ψ at t.

Given $\psi \in \Psi$, $t \in [0, 1]$, let

$$x(t) = \frac{1}{\psi(t)}\,(1-t,\ t)$$

so that $\|x(t)\| = 1$. We identify the dual of $(\underset{\sim}{C}^2, \|\cdot\|)$ with $\underset{\sim}{C}^2$ and denote the dual norm by $\|\cdot\|'$. Clearly $\|\cdot\|' \in N_a$.

Lemma 4. Let $\psi \in \Psi$ and let $\underset{\sim}{C}^2$ be given the corresponding norm.

(i) For $0 < t < 1$, $D(\underset{\sim}{C}^2, x(t)) = \{(\psi(t)-t\gamma,\ \psi(t)+(1-t)\gamma) : \gamma \in G(t)\}$.

(ii) $D(\underset{\sim}{C}^2, x(0)) = \{(1,\ z(1+\gamma)) : \gamma \in G(0),\ |z| = 1\}$,

$D(\underset{\sim}{C}^2, x(1)) = \{(z(1-\gamma),\ 1) : \gamma \in G(1),\ |z| = 1\}$.

Proof. (i) Let $(\alpha, \beta) \in D(\underset{\sim}{C}^2, x(t))$. We show first that $\alpha, \beta \in \underset{\sim}{R}$. We have

$$|\alpha z + \beta w| \leq \|(z, w)\| \qquad (z, w \in \underset{\sim}{C})$$

and since $\|\cdot\|$ is absolute it follows that

$$|(\mathrm{Re}\ \alpha)z + (\mathrm{Re}\ \beta)w| \leq \|(z, w)\| \qquad (z, w \in \underset{\sim}{C}),$$

and hence $(\text{Re}\ \alpha,\ \text{Re}\ \beta) \in D(\underset{\sim}{C}^2,\ x(t))$. It follows from Lemma 2 that $|\text{Re}\ \alpha| = |\alpha|$ or $|\text{Re}\ \beta| = |\beta|$. Since

$$(1 - t)\ \text{Re}\ \alpha + t\ \text{Re}\ \beta = \psi(t) = (1 - t)\alpha + t\beta,$$

we have in fact $\text{Re}\ \alpha = \alpha$, $\text{Re}\ \beta = \beta$. The real line

$$\alpha\xi + \beta\eta = 1$$

is thus a tangent to the curve $x(t)$ $(0 \leq t \leq 1)$ and a simple calculation shows that (α, β) is of the form

$$(\psi(t) - t\gamma,\quad \psi(t) + (1 - t)\gamma)$$

for some $\gamma \in G(t)$. Conversely it is now clear that any such functional is in $D(\underset{\sim}{C}^2,\ x(t))$.

(ii) Let $(\alpha, \beta) \in D(\underset{\sim}{C}^2,\ x(0))$. Then $\alpha = 1$, and since $\|\cdot\|'$ is absolute we have

$$1 - t + t|\beta| \leq \psi(t) \qquad (0 < t < 1).$$

This gives

$$-1 \leq |\beta| - 1 \leq \frac{\psi(t) - 1}{t} \qquad (0 < t < 1),$$

and so $|\beta| = 1 + \gamma$ for some $\gamma \in G(0)$. Since $\|\cdot\|'$ is absolute, it is clear that $(1,\ z(1+\gamma)) \in D(\underset{\sim}{C}^2,\ x(0))$ for $\gamma \in G(0)$, $|z| = 1$. The other part is similar.

Given $\|(z, w)\| = 1$, $|\lambda| = |\mu| = 1$, note that

$$(\alpha, \beta) \in D(\underset{\sim}{C}^2,\ (z, w)) \Longleftrightarrow (\lambda^*\alpha,\ \mu^*\beta) \in D(\underset{\sim}{C}^2,\ (\lambda z, \mu w)).$$

This remark together with Lemma 4 give a complete description of $\Pi(\underset{\sim}{C}^2,\ \|\cdot\|)$ in terms of ψ and its one-sided derivatives.

Remark. McGregor [180] shows that the set $\{(t, \gamma) : t \in [0, 1],\ \gamma \in G(t)\}$ is homeomorphic to $[0, 1]$, in fact, under the mapping

$(t, \gamma) \to \frac{1}{3}(1 + t + \gamma)$.

Let Ξ denote the set $\{(t, \gamma) : t \in [0, 1],\ \gamma \in G(t)\}$, and let ξ, η be defined on Ξ by

$$\xi(t, \gamma) = 1 - \frac{t\gamma}{\psi(t)}, \qquad \eta(t, \gamma) = 1 + \frac{(1-t)\gamma}{\psi(t)} .$$

Lemma 5. Let $T \in B(\underset{\sim}{C}^2, \|\cdot\|)$ have matrix representation

$$T = \begin{bmatrix} a & b \\ c & d \end{bmatrix} .$$

Then

$$V(T) = \{\xi(t, \gamma)[(1 - t)a + tbz] + \eta(t, \gamma)[(1 - t)cz^* + td] : (t, \gamma) \in \Xi,\ |z| = 1\}.$$

Proof. Straightforward computation.

Lemma 5 represents $V(T)$ as the union of a family of ellipses; McGregor [180] has shown that $V(T)$ is actually the union of the convex hulls of these ellipses and thence that $V(T)$ is simply connected, in particular $V(T)$ always contains the line segment $[a, d]$. We give below two examples of numerical ranges due to McGregor [180]; further applications of this section appear in sections 29 and 32.

Example 6. Let $\|(z, w)\| = \max\{\|(z, w)\|_\infty, \frac{3}{\sqrt{10}}\|(z, w)\|_2\}$ and let

$$T = \begin{bmatrix} 0 & \frac{1}{3} \\ \frac{1}{3} & 1 \end{bmatrix}$$

Then $V(T)$ consists of two 'tear-drops' joined by a line segment.

Proof. The corresponding function ψ is given by

$$\psi(t) = \begin{cases} 1 - t & 0 \le t \le \frac{1}{4} \\ \frac{3}{\sqrt{10}}[(1-t)^2 + t^2]^{\frac{1}{2}} & \frac{1}{4} \le t \le \frac{3}{4} \\ t & \frac{3}{4} \le t \le 1 \end{cases}$$

For $t \in [\frac{1}{4}, \frac{3}{4}]$, the corresponding ellipses of $V(T)$ are line segments on the real axis. The remaining ellipses satisfy $\text{Re } z \leq \frac{3}{10}$ or $\text{Re } z \geq \frac{7}{10}$. A sketch of $V(T)$ is given below.

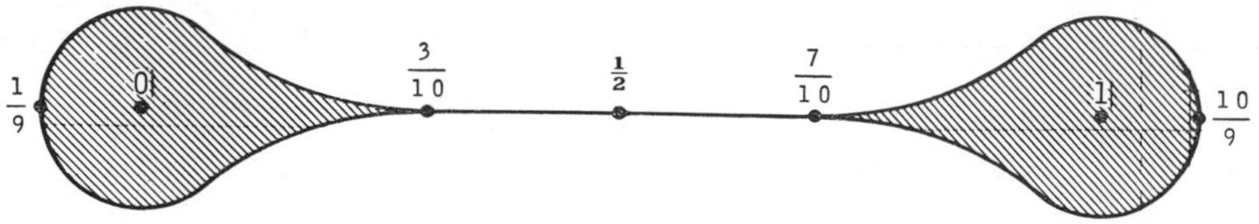

Remark. By chopping off several segments of the l_2-ball one can readily produce a numerical range that looks like a 'string of beads'.

Example 7. Let X be the l_∞ direct sum of $\underset{\sim}{C}^2$ (as in Example 6) and $\underset{\sim}{C}$, and let

$$S = \begin{bmatrix} T & 0 \\ 0 & \frac{1}{2} \end{bmatrix} .$$

Then $V(S)$ is a 'sharp bow-tie' centred on $\frac{1}{2}$.

Proof. By Remark (2) after Lemma 15. 2, we have

$$V(S) = \cup \, \{[\tfrac{1}{2}, \lambda] : \lambda \in V(T)\}.$$

Observe in relation to §20 that $\frac{1}{2}$ is an eigenvalue of S on the boundary of $V(S)$ and there is no open disc in $\underset{\sim}{C} \backslash V(S)$ whose closure contains $\frac{1}{2}$.

6 · Algebra numerical ranges

22. SOME ELEMENTARY OBSERVATIONS ON THE ALGEBRA NUMERICAL RANGE

Let A be a complex unital normed algebra. Given $a \in A$ we recall that the numerical range $V(a)$ of a has the simple description

$$V(a) = \{f(a) : f \in D(A, 1)\},$$

where $D(A, 1) = \{f \in A' : f(1) = 1 = \|f\|\}$. We collect here some simple facts about $V(a)$ none of which were made explicit in Volume I of these notes. The first observation is due to J. P. Williams.

Lemma 1. <u>Given</u> $a \in A$ <u>we have</u>

$$V(a) = \bigcap_{z \in \mathbb{C}} \{\lambda : |z - \lambda| \leq \|z - a\|\}.$$

Proof. Similar to the proof of Lemma 15. 1.

Remark. The lemma exhibits $V(a)$ as an intersection of closed discs.

Lemma 2. <u>Let</u> A, B <u>be complex unital normed algebras and let</u> ϕ <u>be a norm decreasing homomorphism from</u> A <u>to</u> B <u>with</u> $\phi(1) = 1$. <u>Then</u> $V(B, \phi(a)) \subset V(A, a)$.

Proof. Given $\lambda \in V(B, \phi(a))$, choose $g \in D(B, 1)$ with

$$\lambda = g(\phi(a)).$$

Define f on A by

$$f(x) = g(\phi(x)) \quad (x \in A).$$

Then $f(1) = 1$, $|f(x)| \leq \|g\| \, \|\phi\| \, \|x\|$ $(x \in A)$, so that $f \in D(A, 1)$ and $\lambda = f(a) \in V(a)$.

As a special case of Lemma 2 we may take $B = A/J$ where J is a closed two-sided ideal of A; for this case we have a more precise result.

Lemma 3. _Let_ J _be a closed two-sided ideal of_ A _and let_ $\pi : A \to A/J$ _be the canonical mapping. Then for_ $a \in A$ _we have_

$$V(A/J, \pi(a)) = \cap \{V(a + j) : j \in J\}.$$

Proof. Given $z, \lambda \in \mathbb{C}$, we note that

$$|z - \lambda| \leq \|\pi(z - a)\|$$

if and only if

$$|z - \lambda| \leq \|z - (a + j)\| \qquad (j \in J).$$

The result follows from Lemma 1.

Lemma 4. _Let_ $A_{\mathbb{R}}$ _denote the algebra_ A _regarded as a normed algebra over_ $\mathbb{R}$. _Then for_ $a \in A$ _we have_

$$V(A_{\mathbb{R}}, a) = \operatorname{Re} V(A, a).$$

Proof. Apply Lemma 15.3.

Let A, B be unital normed algebras. Then $A \times B$ is a unital normed algebra with the usual pointwise operations and norm $\|\cdot\|_\infty$ defined by

$$\|(a, b)\|_\infty = \max\{\|a\|, \|b\|\}.$$

Lemma 5. _Let_ $(a, b) \in (A \times B, \|\cdot\|_\infty\}$. _Then_

$$V((a, b)) = \{\alpha\lambda + (1-\alpha)\mu : \lambda \in V(a),\ \mu \in V(b),\ \alpha \in [0, 1]\}.$$

Proof. $(A \times B)'$ is isometrically isomorphic to $A' \times B'$ under the correspondence

$$\phi(x, y) = f(x) + g(y) \qquad (x \in A,\ y \in B)$$
$$\|\phi\| = \|f\| + \|g\|.$$

Since 1 is an extreme point of the unit disc, it follows that $D(A \times B, (1,1))$ is the set of functionals ϕ of the form $\phi(x, y) = \alpha f(x) + (1-\alpha)g(y)$, with $f \in D(A, 1)$, $g \in D(B, 1)$, $\alpha \in [0, 1]$.

With A, B as above let $A \otimes B$ be the algebraic tensor product of A and B. The weak tensor product norm λ on $A \otimes B$ is defined by

$$\lambda(x) = \sup\{\,|\Sigma f(x_j)g(y_j)| : x = \Sigma x_j \otimes y_j,\ f \in A',\ g \in B',\ \|f\| \le 1,\ \|g\| \le 1\,\},$$

and the projective tensor product norm p on $A \otimes B$ is defined by

$$p(x) = \inf\{\Sigma \|x_j\|\,\|y_j\| : x = \Sigma x_j \otimes y_j\}.$$

Lemma 6. *Let $A \otimes B$ have the weak or projective tensor product norm. Then*

$$V(a \otimes b) \supset \mathrm{co}\{\lambda\mu : \lambda \in V(a),\ \mu \in V(b)\}.$$

Proof. Given $f \in D(A, 1)$, $g \in D(B, 1)$, let

$$\phi(a \otimes b) = f(a)\, g(b) \qquad (a \in A,\ b \in B).$$

Then ϕ has a well-defined linear extension to $A \otimes B$, and

$$\phi(x) = \Sigma f(x_j)\, g(y_j)$$

for all representations $x = \Sigma x_j \otimes y_j$ of x. Thus

$$|\phi(x)| \le \lambda(x) \le p(x) \qquad (x \in A \otimes B).$$

Since $\phi(1 \otimes 1) = 1$ and $V(a \otimes b)$ is convex, the result follows.

Remark. The above result holds for any cross-norm α on $A \otimes B$ such that $\lambda \le \alpha$.

As an example for which equality holds in Lemma 6, let E, F be compact Hausdorff spaces, and let $A = C(E)$, $B = C(F)$. It is well known ([216]) that the completion of $(A \otimes B, \lambda)$ is isometrically isomorphic to $C(E \times F)$ and so

$$V(a \otimes b) = \mathrm{co}(a(E)\, b(F)) = \mathrm{co}(\mathrm{co}\, a(E)\, \mathrm{co}\, b(F)) = \mathrm{co}\, V(a)\, V(b).$$

As an example for which the inclusion in Lemma 6 is strict, let $G = \underset{\sim}{Z}_2 = \{e, x\}$ and let $A = B = l_1(G)$. It is well known ([216]) that the completion of $(A \otimes B, p)$ is isometrically isomorphic to $l_1(G \times G)$. Let $a = b = e + x$. Then

$$V(a) = V(b) = \{z : |z - 1| \leq 1\}$$
$$V(a \otimes b) = \{z : |z - 1| \leq 3\}$$

and clearly $-2 \notin \mathrm{co}\, V(a)\, V(b)$.

23. MAPPING THEOREMS FOR NUMERICAL RANGES

Some mapping theorems for numerical ranges were given in Volume I of these notes. For example, NRI Theorem 4.8 may be rephrased as:

$$V(a) \subset \{z : |z| \leq 1\} \Rightarrow V(a^n) \subset \{z : |z| \leq n!\,(e/n)^n\}.$$

It will be shown in the next section that this inclusion is in fact best possible in general. In particular the power inequality

$$v(a^n) \leq v(a)^n$$

fails in general. A further negative result even for Hermitian elements was given in NRI Example 6.1 where we have

$$V(h) = [-1, 1], \quad V(h^2) \not\subset [0, 1].$$

It is thus clear that strong mapping theorems for numerical ranges can be obtained only under restrictive hypotheses. We give some of these results in this section.

We recall that the spectrum of an element behaves very well under mappings, in particular we have

$$Sp(p(a)) = p(Sp(a)) \qquad (p \text{ a polynomial}).$$

This mapping theorem gives a partial characterization of $Sp(\cdot)$.

Theorem 1. Let A be a complex unital Banach algebra. Let Φ be a mapping from A into the compact subsets of $\underset{\sim}{C}$ such that

(i) Φ is upper semi-continuous,

(ii) $Sp(A, a) \subset \Phi(a) \qquad (a \in A)$,

(iii) $\Phi(p(a)) = p(\Phi(a)) \qquad (a \in A, \ p$ a polynomial).

Then for each $a \in A$, $\Phi(a) \subset Sp(A(a), a)$ where $A(a)$ is the closed algebra generated by 1 and a.

Proof. Let

$$\phi(a) = \sup\{ |\lambda| : \lambda \in \Phi(a) \} \qquad (a \in A).$$

Then by (iii), $\phi(0) = 0$. There exists $M > 0$ such that

$$\phi(a) \leq M\|a\| \qquad (a \in A).$$

Otherwise we may choose $\{a_n\}$ such that $a_n \to 0$, $\phi(a_n) \to 1$ and this contradicts the upper semi-continuity of Φ. Let $a \in A$, $z \in \Phi(a)$. If a is an algebraic element let p be the corresponding minimal polynomial, and then $p(\Phi(a)) = \{0\}$. It follows that $\Phi(a) \subset Sp(A, a)$. If there is no such polynomial, let F be defined on the algebra generated by 1 and a by

$$F(p(a)) = p(z) \qquad (p \text{ a polynomial}).$$

Then

$$|F(p(a))| \leq \sup\{ |p(\lambda)| : \lambda \in \Phi(a) \} = \phi(p(a)) \leq M\|p(a)\|$$

so that F extends to a homomorphism on $A(a)$ and therefore $z \in Sp(A(a), a)$.

Since the numerical range is compact convex valued on A we wish also to have a corresponding characterization of co Sp(·).

Theorem 2. Let A be a complex unital Banach algebra. Let Ψ be a mapping from A into the compact convex subsets of $\underset{\sim}{C}$ such that

(i) Ψ is upper semi-continuous,

(ii) $Sp(a) \subset \Psi(a) \quad (a \in A)$,

(iii) $\Psi(p(a)) \subset \text{co } p(\Psi(a)) \quad (a \in A, \ p$ a polynomial),

(iv) $\psi(a + b) \leq \psi(a) + \psi(b) \quad (a, b \in A, \ ab = ba)$,

where $\psi(a) = \sup\{|\lambda| : \lambda \in \Psi(a)\}$. Then $\Psi(a) = \text{co } Sp(a) \quad (a \in A)$.

Proof. It is sufficient to show that $\psi(a) \leq \rho(a) \ (a \in A)$. As in Theorem 1, there is $M > 0$ such that $\psi(a) \leq M\|a\| \ (a \in A)$, and clearly by (iv), ψ is a semi-norm on A(a). We have $z \in \Psi(a)$ if and only if

$$|\lambda + \mu z| \leq \psi(\lambda + \mu a) \qquad (\lambda, \mu \in \underset{\sim}{C}) .$$

The Hahn-Banach theorem now applies to give

$$\Psi(a) = \{f(a) : f \in (A(a))', \ f(1) = 1, \ |f(x)| \leq \psi(x) \ (x \in A(a))\}. \qquad (1)$$

Let $a \in A$. By (iii), we have $\psi(a^2) \leq (\psi(a))^2$. Suppose that $\psi(a^2) < (\psi(a))^2$. Then $a \neq 0$, and since $\Psi(\zeta a) = \zeta\Psi(a) \ (\zeta \in \underset{\sim}{C})$, we may suppose that $\psi(a) = 1$ and $-1 \in \Psi(a)$. By (1), we may choose $f \in (A(a))'$ with $f(1) = 1$, $|f(x)| \leq \psi(x) \ (x \in A(a))$, and $f(a) = -1$. Since

$$\text{Re } f(a^2) \leq |f(a^2)| \leq \psi(a^2) < 1,$$

this gives $\text{Re } f((1 + \tfrac{1}{2}a)^2) < \tfrac{1}{4}$. Therefore, by (1), there exists $\alpha \in \Psi((1 + \tfrac{1}{2}a)^2)$, with $\text{Re } \alpha < \tfrac{1}{4}$. But

$$\Psi(1 + \tfrac{1}{2}a) \subset \{1 + z : |z| \leq \tfrac{1}{2}\},$$

and so, by (iii),

$$\Psi((1 + \tfrac{1}{2}a)^2) \subset \text{co}\{(1 + z)^2 : |z| \leq \tfrac{1}{2}\}.$$

But an elementary calculation gives

$$\mathrm{Re}((1+z)^2) = 1 + 2r\cos\theta + r^2\cos 2\theta \quad (z=re^{i\theta},\ 0\le r\le\tfrac{1}{2},\ \theta\in\mathbb{R}),$$

which gives the contradictory conclusion $\mathrm{Re}\,\alpha \ge \frac{1}{4}$. This proves that $\psi(a^2) = (\psi(a))^2$ $(a \in A)$. Therefore

$$\psi(a) = (\psi(a^{2^n}))^{2^{-n}} \le M^{2^{-n}} \|a^{2^n}\|^{2^{-n}} \qquad (n = 1, 2, \ldots),$$

and so $\psi(a) \le \rho(a)$, as required.

Remarks. (1) The above proof does not use the full strength of condition (iii). Only polynomials of degree 1 and the polynomial z^2 were used.

(2) We do not know if condition (iv) is necessary.

We have already seen in NRI Theorem 4.7 that the condition $v = \rho$ is a very restrictive condition on a complex unital Banach algebra. The next result gives a 'functional calculus' characterization of such algebras.

Theorem 3. Let A be a complex unital Banach algebra. Then the following are equivalent.

(i) $V(a) = \mathrm{co}\,\mathrm{Sp}(a)$ $(a \in A)$,

(ii) $V(a^2) \subset \mathrm{co}\{z^2 : z \in V(a)\}$.

Proof. (i) $\Rightarrow$ (ii) is elementary, and (ii) $\Rightarrow$ (i) follows from Theorem 2 and the above Remark (1).

Remark. One can show that the even stronger condition

$$V(a^2) = \mathrm{co}\{z^2 : z \in V(a)\} \qquad (a \in A)$$

forces A to be $\mathbb{C}$. The proof, although a little complicated, depends only on consideration of complex polynomials as mappings of $\mathbb{C}$.

Lemma 4. Let A be a complex unital Banach algebra. Then the following are equivalent.

(i) $V(a) = \mathrm{co}\,\mathrm{Sp}(a)$ $(a \in A)$,

(ii) $\|\exp(a)\| = \rho(\exp(a))$ $(a \in A)$.

Proof. This is a simple consequence of NRI Theorems 3.4 and 3.8.

Theorem 5. Let A be a complex unital Banach algebra with $v = \rho$. Then $\|a\| \leq \frac{1}{2}e\rho(a)$ $(a \in A)$.

Proof. Let $a \in A$ with $\rho(a) < 1$ and let Γ be the unit circle. Then

$$\begin{aligned}
&\frac{1}{2\pi i}\int_\Gamma \exp((za+1)(za-1)^{-1})\frac{dz}{z^2} \\
&= \frac{1}{2\pi i}\int_\Gamma \frac{1}{2\pi i}\int_\Gamma \exp\left(\frac{zw+1}{zw-1}\right)(w-a)^{-1}\,dw\,\frac{dz}{z^2} \\
&= \frac{1}{2\pi i}\int_\Gamma \frac{1}{2\pi i}\int_\Gamma \exp\left(\frac{zw+1}{zw-1}\right)\frac{dz}{z^2}(w-a)^{-1}\,dw \\
&= \frac{1}{2\pi i}\int_\Gamma -\frac{2w}{e}(w-a)^{-1}dw \\
&= -\frac{2}{e}a.
\end{aligned}$$

Given $z \in \Gamma$, $\lambda \in Sp((za+1)(za-1)^{-1})$ we have $\operatorname{Re}\lambda < 0$, and so, by Lemma 4,

$$\|\exp((za+1)(za-1)^{-1})\| \leq 1 \qquad (z \in \Gamma).$$

Therefore $\frac{2}{e}\|a\| \leq 1$ and the result follows.

Corollary 6. Let $v = \rho$ in A, let $a \in A$ with $v(a) = 1$, and let p be a polynomial that maps the unit disc into itself. Then $\|p(a)\| \leq \frac{1}{2}e$.

Proof. Since $Sp(p(a)) = p(Sp(a))$ we have $\rho(p(a)) \leq 1$. The result now follows from the theorem.

We show in §25 that Theorem 5 and its corollary are best possible for Banach algebras with $v = \rho$. For B*-algebras the numerical range admits a powerful functional calculus as given in the next two theorems which are due to Kato [38] and Berger and Stampfli [5].

Let A be a B*-algebra and let $a \in A$. We write $\operatorname{Re} a = \frac{1}{2}(a + a^*)$ and we recall that Re a is Hermitian. For the remainder of this section we shall use the notation

$$\Delta = \{z \in \underset{\sim}{C} : |z| \le 1\}, \qquad P = \{z \in \underset{\sim}{C} : \mathrm{Re}\, z \ge 0\}.$$

Lemma 7. Let A be a B^*-algebra and let $a \in A$. Then $v(a) \le 1$ if and only if $\mathrm{Re}(1 - za)^{-1} \ge 0$ $(|z| < 1)$.

Proof. We have the following chain of equivalences:

$$\begin{aligned} v(a) \le 1 &\iff |f(a)| \le 1 \quad (f \in D(1)) \\ &\iff \mathrm{Re}\, f(1 - za) \ge 0 \quad (f \in D(1), \quad |z| < 1) \\ &\iff f(\mathrm{Re}(1 - za)) \ge 0 \quad (f \in D(1), \quad |z| < 1) \\ &\iff \mathrm{Re}(1 - za) \ge 0 \quad (|z| < 1) \\ &\iff \mathrm{Re}((1 - za)^{-1}) \ge 0 \quad (|z| < 1). \end{aligned}$$

Theorem 8. Let A be a B^*-algebra, let $a \in A$ with $v(a) \le 1$ and let F be analytic on a neighbourhood of Δ with $F(\Delta) \subset P$. Then $V(F(a)) \subset P - \mathrm{Re}\, F(0)$.

Proof. For $|z| < 1$ we have

$$F(z) = -F(0)^* + \frac{1}{\pi} \int_0^{2\pi} (\mathrm{Re}\, F(e^{it}))(1 - ze^{-it})^{-1} dt.$$

Given $0 < r < 1$, we may now use the Fubini argument in the proof of Theorem 5 to obtain

$$F(ra) = -F(0)^* 1 + \frac{1}{\pi} \int_0^{2\pi} (\mathrm{Re}\, F(e^{it}))(1 - rae^{-it})^{-1} dt.$$

Lemma 7 now gives

$$\mathrm{Re}\, F(ra) + \mathrm{Re}\, F(0) 1 \ge 0 \qquad (0 < r < 1),$$

and this gives the required conclusion.

Theorem 9. Let A be a B^*-algebra, let $a \in A$ with $v(a) \le 1$, let F be analytic on a neighbourhood of Δ, and let $F(0) = 0$, $F(\Delta) \subset \Delta$. Then $v(F(a)) \le 1$.

Proof. Given $|z| < 1$, let G be defined on a neighbourhood of Δ by

$$G(w) = \frac{1 + zF(w)}{1 - zF(w)} .$$

Then $G(\Delta) \subset P$, $G(0) = 1$, and so Theorem 8 gives

$$V(G(a)) \subset P - 1, \qquad \mathrm{Re}(G(a) + 1) \geq 0.$$

Thus

$$\mathrm{Re}((1 - zF(a))^{-1}) \geq 0 \qquad (|z| < 1),$$

and so $v(F(a)) \leq 1$ by Lemma 7.

Corollary 10. <u>The power inequality holds for B^*-algebras.</u>

Remarks. (1) Let F be as in Theorem 9, let $\lambda, \mu \in \Delta$ and let $|\lambda| + |\mu| \leq 1$. By applying the result of Theorem 9 to shift operators on Hilbert spaces of dimensions 2 and 3, we may obtain the inequalities

$$|F(\lambda)| + |\mu F'(\lambda)| \leq 1, \qquad |F(\lambda)| + |\frac{\mu^2}{2} F''(\lambda)| \leq 1.$$

(2) It would be of interest to characterize those Banach algebras for which the conclusion of Theorem 9 holds.

Theorems 8 and 9 give a numerical range functional calculus for a special class of Banach algebras. In another direction we may seek functional calculus results for special elements of Banach algebras. We consider below the case of idempotent elements.

Given $j \in A$, $j^2 = j$, $(j \neq 0,\ j \neq 1)$, the algebra generated by 1 and j may be identified with $\underset{\sim}{C}^2$ under pointwise multiplication where

$$j = (1, 0), \qquad 1 - j = (0, 1).$$

The results below may thus also be considered as a study of algebra norms on $\underset{\sim}{C}^2$; the results appear in [135].

Theorem 11. <u>Let</u> A <u>be a complex unital Banach algebra and let</u> j <u>be a non-zero idempotent of</u> A <u>with</u> $\max \mathrm{Re}\, V(j) \leq 1$. <u>Then</u> $\|j\| = 1$. <u>In particular,</u> $j^2 = j$, $v(j) = 1$ <u>implies</u> $\|j\| = 1$.

Proof. For $t > 0$ we have $\|\exp(tj)\| \le e^t$ and so

$$\|1 + (e^t - 1)j\| \le e^t$$

$$\|j\| \le \frac{e^t + 1}{e^t - 1} \qquad (t > 0).$$

Therefore $\|j\| \le 1$. Since $j \ne 0$, we have $1 \in Sp(j)$ and hence $\|j\| = 1$.

Theorem 12. Let j be a non-zero idempotent in a complex unital Banach algebra A. Then $\|j\| < e\, v(j)$.

Proof. We may assume by Theorem 11 that $v(j) > 1$. We have

$$\|\exp(tj)\| \le e^{|t| v(j)} \qquad (t \in \underset{\sim}{R})$$

and so

$$\|\tfrac{1}{2}\exp(tj) - \tfrac{1}{2}\exp(-tj)\| \le e^{tv(j)} \qquad (t > 0).$$

It follows that

$$\|j\| \le \frac{e^{tv(j)}}{\sinh t} \qquad (t > 0),$$

and we may minimize the right-hand side to give

$$\frac{\|j\|}{\alpha} \le (1 + \frac{1}{\alpha})(1 + \frac{2}{\alpha - 1})^{\frac{1}{2}(\alpha - 1)},$$

where $\alpha = v(j)$. The function of α on the right-hand side is strictly increasing on $(1, \infty)$ with limit e at infinity. The proof is complete.

It is shown in [135] that for each $t \in (e^{-1}, 1)$ there is a complex unital Banach algebra A and an idempotent j in A such that $v(j) = t\|j\|$.

Consider again the case $\|j\| = 1$. The following results are given in [135]. The extremal case $V(j) = \Delta$ is attained. Now let $\|1 - j\| = 1$, which corresponds to

$$\|(x, y)\| = \max(x, y) \qquad (x, y \ge 0).$$

This gives $V(j) \subset \Delta \cap \{z : |z - 1| \le 1\}$ and the extremal case is attained. Suppose further that $\|j - (1 - j)\| = 1$, which corresponds to

$$\|(x, y)\| = \max(|x|, |y|) \qquad (x, y \in \mathbb{R}).$$

This gives $V(j) \subset \{z : |z - \frac{1}{2}| \leq \frac{1}{2}\}$ and the extremal case is again attained.

24. THE EXTREMAL ALGEBRA Ea(K)

In NRI Theorem 4.8 we gave estimates for the norms of powers of an element in a complex unital Banach algebra in terms of the corresponding numerical radius. We show here that these estimates are best possible; indeed there is a complex unital Banach algebra A and a non-zero $u \in A$ such that

$$\|u^n\| = n!\ (e/n)^n\ v(u)^n \qquad (n = 1, 2, 3, \ldots).$$

This result appears to have been first proved by Browder [125] and also independently by Crabb [132] and Bollobás [116].

The normalizing condition $v(u) = 1$ is equivalent to the condition that $V(u) \subset K$, where K is the closed unit disc. More generally, Bollobás [117] has constructed the algebra generated by u with the largest norm for which $V(u) \subset K$, where K is an arbitrary non-empty compact convex subset of $\mathbb{C}$. The construction of Bollobás [117] is given in terms of formal power series and requires only elementary techniques. The subsequent construction of Crabb, Duncan and McGregor [134] is less elementary, but we shall follow their construction since it exhibits the extremal algebras in terms of classical algebras of analytic functions.

Notation. Let K be a convex subset of $\mathbb{C}$ with at least two points. We assume that K is normalized so that $0, 1 \in K$, $\sup\{|z| : z \in K\} = 1$, and either $0 \in \mathrm{int}(K)$ or $K \subset \mathbb{R}$.

We denote by u the function defined on K by

$$u(z) = z \qquad (z \in K),$$

and by ω the function defined on $\mathbb{C}$ by

$$\omega(\lambda) = \sup\{|e^{\lambda z}| : z \in K\}.$$

We note that

$$\omega(\lambda) = \|\exp(\lambda u)\|_\infty \qquad (\lambda \in \underset{\sim}{C}),$$

where $\|\cdot\|_\infty$ is the supremum norm for functions on K, and $\exp(\lambda u)$ denotes the function $z \to e^{\lambda z}$ $(z \in K)$.

We denote by $M(\underset{\sim}{C})$ the Banach space of all (finite) complex regular Borel measures on $\underset{\sim}{C}$, and by $M^\omega(\underset{\sim}{C})$ the set of all $\mu \in M(\underset{\sim}{C})$ such that

$$\|\mu\|_\omega = \int \omega \, d|\mu| < \infty.$$

Given $\mu \in M^\omega(\underset{\sim}{C})$, we define a function f_μ on K by

$$f_\mu(z) = \int e^{z\lambda} \, d\mu(\lambda) \qquad (z \in K),$$

where the integral exists because $|e^{z\lambda}| \le \omega(\lambda)$. We denote by $Ea(K)$ the set of all functions f_μ with $\mu \in M^\omega(\underset{\sim}{C})$. We note that many different measures μ in $M^\omega(\underset{\sim}{C})$ may give the same function f_μ on K, and for each $f \in Ea(K)$ we define $\|f\|$ by

$$\|f\| = \inf\{\|\mu\|_\omega : \mu \in M^\omega(\underset{\sim}{C}), \ f_\mu = f\}.$$

Lemma 1. Let $\zeta \in \underset{\sim}{C}$, let Γ be any circle with ζ in its interior, let μ be the measure on Γ given by $d\mu(\lambda) = \frac{1}{2\pi i}\frac{1}{\lambda-\zeta} d\lambda$, and let ν be the measure with unit mass concentrated at ζ. Then $\mu, \nu \in M^\omega(\underset{\sim}{C})$ and

$$\exp(\zeta u) = f_\mu = f_\nu.$$

Proof. $M^\omega(\underset{\sim}{C})$ contains all elements of $M(\underset{\sim}{C})$ with compact supports, and so $\mu \in M^\omega(\underset{\sim}{C})$. Now apply Cauchy's integral formula. That $f_\nu = \exp(\zeta u)$ is even more obvious.

Theorem 2. $(Ea(K), \|\cdot\|)$ is a complex unital Banach algebra of continuous complex valued functions on K, and

(i) $\|f\|_\infty \le \|f\|$ $(f \in Ea(K))$,

(ii) $\|\exp(\zeta u)\|_\infty = \|\exp(\zeta u)\|$ $(\zeta \in \underset{\sim}{C})$,

(iii) $u \in Ea(K)$, and for $\zeta \in \underset{\sim}{C}$ the function $\exp(\zeta u)$ is the exponential of ζu in the Banach algebra sense,

(iv) $Ea(K)$ is the (norm) closed linear hull of $\{\exp(\zeta u) : \zeta \in \underset{\sim}{C}\}$ and so $Ea(K)$ is generated by u.

Proof. We have

$$\omega(\zeta_1 + \zeta_2) = \|\exp(\zeta_1 u)\exp(\zeta_2 u)\|_\infty \le \omega(\zeta_1)\,\omega(\zeta_2) \qquad (\zeta_1, \zeta_2 \in \underset{\sim}{C}),$$

and therefore, given $\mu, \nu \in M^\omega(\underset{\sim}{C})$,

$$\iint \omega(\alpha + \beta)\, d|\mu|(\alpha)\, d|\nu|(\beta) \le \|\mu\|_\omega \|\nu\|_\omega .$$

This shows that $\mu * \nu \in M^\omega(\underset{\sim}{C})$, and

$$\|\mu * \nu\|_\omega \le \|\mu\|_\omega \|\nu\|_\omega .$$

We have

$$\begin{aligned} f_\mu(z)\, f_\nu(z) &= \iint e^{z(\alpha+\beta)}\, d\mu(\alpha)\, d\nu(\beta) \\ &= \int e^{z\lambda}\, d(\mu * \nu)(\lambda) \\ &= f_{\mu*\nu}(z). \end{aligned}$$

Therefore $Ea(K)$ is a subalgebra of $C(K)$, and

$$\|fg\| \le \|f\|\,\|g\| \qquad (f, g \in Ea(K)).$$

We have, for $z \in K$ and $\mu \in M^\omega(\underset{\sim}{C})$, the inequality

$$|f_\mu(z)| = \left|\int e^{z\lambda} d\mu(\lambda)\right| \le \int \|\exp \lambda u\|_\infty d|\mu|(\lambda) = \int \omega d|\mu| = \|\mu\|_\omega,$$

which proves (i).

Given $\zeta \in \underset{\sim}{C}$, we have $\exp(\zeta u) = f_\nu$ with ν the measure described in Lemma 1. Then

$$\|\exp(\zeta u)\| \le \|\nu\|_\omega = \int \omega(\lambda)\, d(\nu)\,(\lambda) = \omega(\zeta),$$

which proves (ii). In particular (ii) holds with $\zeta = 0$ and gives $\|1\| = 1$. Thus Ea(K) is a complex unital normed algebra.

To show that Ea(K) is complete, it is enough to show that every absolutely convergent series in Ea(K) converges. Let $f_n \in \mathrm{Ea}(K)$ with $\sum_1^\infty \|f_n\| < \infty$, and choose $\mu_n \in M^\omega(\underset{\sim}{C})$ with

$$f_{\mu_n} = f_n, \quad \|\mu_n\|_\omega \leq \|f_n\| + 2^{-n} \qquad (n = 1, 2, \ldots).$$

Let $\sigma_n = \sum_{k=1}^n \mu_k$, $\sigma = \sum_{k=1}^\infty \mu_k$, $s_n = \sum_{k=1}^n f_k$. By (i), $\sum_{k=1}^\infty f_k(z)$ converges uniformly on K to a function s(z), and we have $s_n = f_{\sigma_n}$, $\sigma \in M^\omega(\underset{\sim}{C})$, $s = f_\sigma \in \mathrm{Ea}(K)$. Then $s - s_n = f_{\sigma-\sigma_n}$, and so

$$\|s - s_n\| \leq \|\sigma - \sigma_n\|_\omega \to 0.$$

Thus Ea(K) is complete.

To prove (iii), take Γ to be the unit circle and note that the formula

$$z = \frac{1}{2\pi i} \int e^{z\lambda} \frac{1}{\lambda^2}\, d\lambda$$

expresses u in the form $u = f_\mu$ where μ is a measure with compact support and is therefore in $M^\omega(\underset{\sim}{C})$. Given $\zeta \in \underset{\sim}{C}$ let

$$v = \sum_{n=0}^\infty \frac{1}{n!} (\zeta u)^n.$$

Then

$$\begin{aligned} \left|v(z) - \sum_{r=0}^n \frac{1}{r!} (\zeta z)^r\right| &\leq \left\|v - \sum_{r=0}^n \frac{1}{r!} (\zeta u)^r\right\|_\infty \\ &\leq \left\|v - \sum_{r=0}^n \frac{1}{r!} (\zeta u)^r\right\| \\ &\to 0, \text{ as } n \to \infty; \end{aligned}$$

and therefore $v = \exp(\zeta u)$.

It follows from (iii) that the mapping $\zeta \to \exp(\zeta u)$ is analytic and so continuous on $\underset{\sim}{C}$. Therefore for arbitrary $\mu \in M^\omega(\underset{\sim}{C})$

$$f_\mu = \int \exp(\lambda u)\, d\mu(\lambda),$$

where the integral is interpreted as the integral of the Banach space valued function $\exp(\lambda u)$. It follows at once that f_μ is the norm limit of linear combinations of the elements $\exp(\lambda u)$ with $\lambda \in \underset{\sim}{C}$. To complete the proof of (iv) we note that the closed subalgebra generated by u contains $\exp(\lambda u)$ for each $\lambda \in \underset{\sim}{C}$ and so is $Ea(K)$.

Remark. The Banach algebra $Ea(K)$ constructed above is called the <u>extremal algebra for K.</u>

Definition 3. We denote by $D(K)$ the set of all entire functions ϕ such that

$$\|\phi\| = \sup\{(\omega(\lambda))^{-1}|\phi(\lambda)| : \lambda \in \underset{\sim}{C}\} < \infty.$$

It is plain that $(D(K), \|\cdot\|)$ is a Banach space, and our next aim is to prove that it can be identified with the dual space of $Ea(K)$.

Lemma 4. <u>Let</u> $\phi \in D(K)$ <u>and</u> $\mu \in M^\omega(\underset{\sim}{C})$. <u>If</u> $f_\mu = 0$, <u>then</u>

$$\int \phi d\mu = 0.$$

Proof. We assume that $\|\phi\| \le 1$ and that $f_\mu = 0$. We recall that K was normalized so that $0, 1 \in K$, $\sup\{|z| : z \in K\} = 1$, and either $0 \in \text{int } K$ or $K \subset \underset{\sim}{R}$.

Case 1. $0 \in \text{int } K$. Let $\delta > 0$ be chosen so that $\Delta = \{z : |z| < \delta\} \subset K$, and let $E = \text{co}\{1, \Delta\}$. Then $EK \subset K$, and so

$$\omega(z\lambda) \le \omega(\lambda) \qquad (z \in E,\ \lambda \in \underset{\sim}{C}).$$

Therefore

$$|\phi(z\lambda)| \le \omega(z\lambda) \le \omega(\lambda) \qquad (z \in E,\ \lambda \in \underset{\sim}{C}),$$

and so a function G continuous on E and analytic on $\text{int } E$ is defined by

$$G(z) = \int \phi(z\lambda)\, d\mu(\lambda) \qquad (z \in E).$$

Since $\omega(\lambda) \geq e^{\delta|\lambda|}$, λ^n is integrable with respect to μ, and dominated convergence gives

$$\sum_{k=0}^{\infty} \frac{z^n}{n!} \int \lambda^n d\mu(\lambda) = \int e^{z\lambda} d\mu(\lambda) = f_\mu(z) = 0 \qquad (z \in \Delta).$$

Since Δ is a neighbourhood of 0, this gives

$$\int \lambda^n d\mu(\lambda) = 0 \qquad (n = 0, 1, 2, \ldots). \tag{1}$$

Let the power series representation of ϕ be

$$\phi(\lambda) = \sum_{n=0}^{\infty} c_n \lambda^n \qquad (\lambda \in \underset{\sim}{C}).$$

We have

$$|\phi(\lambda)| \leq \omega(\lambda) = \sup\{|e^{\lambda z}| : z \in K\} \leq e^{|\lambda|} \qquad (\lambda \in \underset{\sim}{C}),$$

and so the Cauchy estimates give

$$|c_n| \leq r^{-n} e^r \qquad (r > 0).$$

Therefore $|c_n| \leq (e/n)^n$, and

$$\sum_{n=0}^{\infty} |c_n \lambda^n| \leq e^{e|\lambda|} \qquad (\lambda \in \underset{\sim}{C}).$$

Since also $\omega(\lambda) \geq e^{\delta|\lambda|}$, it follows by dominated convergence that

$$G(z) = \sum_{n=0}^{\infty} c_n z^n \int \lambda^n d\mu(\lambda) \qquad (|z| < \delta e^{-1}).$$

By (1), we now have

$$G(z) = 0 \quad (|z| < \delta e^{-1}),$$

analytic continuation gives $G(z) = 0$ $(z \in \operatorname{int} E)$, and continuity gives $G(1) = 0$ as required.

Case 2. $K \subset \underset{\sim}{R}$.

Suppose first that $\phi \in D(K) \cap L^2(i\underset{\sim}{R})$. Since

$$|\phi(\lambda)| \le \|\phi\| \, \omega(\lambda) \qquad (\lambda \in \underset{\sim}{C}),$$

the Paley-Wiener theorem (see for example [174] p. 387) gives the existence of $\rho \in L^2(K)$ such that

$$\phi(\lambda) = \int_K e^{\lambda t} \rho(t) dt \qquad (\lambda \in \underset{\sim}{C}).$$

Then, by Fubini's theorem,

$$\int \phi(\lambda) \, d\mu(\lambda) = \int_K \rho(t) \, \{ \int e^{t\lambda} d\mu(\lambda) \} dt = 0.$$

We suppose now that ϕ is an arbitrary element of $D(K)$, and define ϕ_n for $n = 1, 2, \ldots$ by

$$\phi_n(\lambda) = \phi((1 - \tfrac{1}{n})\lambda) \, \frac{\exp(\lambda/n) - 1}{\lambda/n} \qquad (\lambda \in \underset{\sim}{C}).$$

To complete the proof it is now enough to prove that $\phi_n \in D(K) \cap L^2(i\underset{\sim}{R})$ and that $\phi_n(\lambda) \to \phi(\lambda)$ dominatedly as $n \to \infty$. This follows from the inequalities

$$|\phi_n(\lambda)| \le \begin{cases} e\omega(\lambda) & (\lambda \in \underset{\sim}{C}, \ |\lambda| \le n) \\ 2\omega(\lambda) & (\lambda \in \underset{\sim}{C}, \ |\lambda| > n), \end{cases}$$

$$|\phi_n(it)| \le 2n/|t| \qquad (t \in \underset{\sim}{R}, \ |t| \ge 1).$$

Notation. Given $\phi \in D(K)$ and $f \in Ea(K)$, define $\Phi_\phi(f)$ by

$$\Phi_\phi(f) = \int \phi(\lambda) \, d\mu(\lambda) \qquad (\mu \in M^\omega(\underset{\sim}{C}), \ f_\mu = f).$$

By Lemma 4, $\Phi_\phi(f)$ is well defined (i. e. independent of the choice of μ with $f_\mu = f$). Clearly Φ_ϕ is a linear functional on $Ea(K)$.

Theorem 5. The mapping $\phi \to \Phi_\phi$ is an isometric isomorphism of $D(K)$ onto $(Ea(K))'$, and

$$\phi(\lambda) = \Phi_\phi(\exp \lambda u) \qquad (\lambda \in \underset{\sim}{C}, \ \phi \in D(K)).$$

Proof. With $f_\mu = f$, we have

$$|\Phi_\phi(f)| \le \int \|\phi\| \, \omega(\lambda) \, d|\mu|(\lambda) \le \|\phi\| . \|\mu\|_\omega,$$

and so, by definition of the norm on Ea(K),

$$|\Phi_\phi(f)| \le \|\phi\| \, \|f\|.$$

By Lemma 1, $\exp(\zeta u) = f_\nu$ where ν is the measure with unit mass concentrated at ζ. Therefore

$$\Phi_\phi(\exp \zeta u) = \int \phi(\lambda) \, d\nu(\lambda) = \phi(\zeta) \qquad (\zeta \in \underset{\sim}{C}).$$

Then, by Theorem 2 (ii),

$$\omega(\lambda)^{-1} \, |\phi(\lambda)| \le \omega(\lambda)^{-1} \, \|\Phi_\phi\| \, \|\exp \lambda u\| = \|\Phi_\phi\|.$$

Thus $\|\Phi_\phi\| = \|\phi\|$.

It remains to prove that the range of the mapping is the whole of $(Ea(K))'$. Let $\Phi \in (Ea(K))'$ and define ϕ by

$$\phi(\lambda) = \Phi(\exp(\lambda u)) \qquad (\lambda \in \underset{\sim}{C}).$$

Clearly $\phi \in D(K)$, and so

$$\Phi(\exp(\lambda u)) = \Phi_\phi(\exp(\lambda u)) \qquad (\lambda \in \underset{\sim}{C}).$$

Therefore, by Theorem 2(iv), $\Phi = \Phi_\phi$.

Lemma 6. Let a be an element of a complex unital Banach algebra A. Then $V(A, a) \subset K$ if and only if

$$\|\exp(\lambda a)\| \le \omega(\lambda) \qquad (\lambda \in \underset{\sim}{C}).$$

Proof. By NRI Theorem 3.4, for $\zeta \in \underset{\sim}{C}$,

$$\sup \operatorname{Re}(\zeta V(a)) = \sup \operatorname{Re} V(\zeta a) = \sup \{\frac{1}{r} \log \|\exp(r\zeta a)\| : r > 0\}.$$

Also

$$\frac{1}{r}\log \omega(r\zeta) = \sup\{\mathrm{Re}(\zeta z) : z \in K\}.$$

Corollary 7. $\mathrm{Sp}(u) = V(\mathrm{Ea}(K), u) = K$.

Proof. The evaluation functionals at points of K are non-zero multiplicative linear functionals on $\mathrm{Ea}(K)$. Therefore $K \subset \mathrm{Sp}(u)$. Also, by Lemma 6,

$$\mathrm{Sp}(u) \subset V(\mathrm{Ea}(K), u) \subset K.$$

Corollary 8. The maximal ideal space of $\mathrm{Ea}(K)$ can be identified with K.

Proof. Theorem 2 (iv) and Corollary 7.

Notation. A pair (A, a), with A a complex unital Banach algebra and a an element of A, is said to be subordinate to K if $V(A, a) \subset K$.

If (A, a) is subordinate to K and G is a complex function analytic on a neighbourhood of K, the functional calculus gives a well defined element $G(a)$ of A, since $\mathrm{Sp}(a) \subset K$.

We now give the main theorem of this section.

Theorem 9. (Bollobás [117].) $(\mathrm{Ea}(K), u)$ is subordinate to K, and for every complex function G analytic on a neighbourhood of K and every (A, a) subordinate to K,

(i) $\|G(a)\| \leq \|G(u)\| = \sup\{|\Phi_\phi(G(u))| : \phi \in D(K), \|\phi\| = 1\}$,

(ii) $V(G(a)) \subset V(\mathrm{Ea}(K), G(u)) = \{\Phi_\phi(G(u)) : \phi \in D(K), \phi(0) = \|\phi\| = 1\}$.

Proof. Corollary 7 shows that $(\mathrm{Ea}(K), u)$ is subordinate to K. The equality in (i) follows at once from Theorem 5 and the Hahn-Banach theorem. Also by Theorem 5,

$$\Phi_\phi(1) = \Phi_\phi(\exp 0u) = \phi(0),$$

and the equality in (ii) follows.

Let (A, a) be subordinate to K, let $g \in A'$ with $\|g\| = 1$, and define ϕ on $\underset{\sim}{C}$ by

$$\phi(\lambda) = g(\exp(\lambda a)).$$

Since $V(A, a) \subset K$, Lemma 6 gives

$$|\phi(\lambda)| \leq \|\exp(\lambda a)\| \leq \omega(\lambda) \qquad (\lambda \in \underset{\sim}{C}),$$

and so $\phi \in D(K)$, and $\|\phi\| \leq 1$.

Given a polynomial $P(\lambda) = \alpha_0 + \alpha_1\lambda + \ldots + \alpha_n\lambda^n$ $(\lambda \in \underset{\sim}{C})$, take

$$Q(\lambda) = \alpha_0 + 1!\,\alpha_1\lambda + 2!\,\alpha_2\lambda^2 + \ldots + n!\,\alpha_n\lambda^n,$$

and let μ_P be the measure defined on a simple closed contour Γ containing K by

$$d\mu_P(\lambda) = \frac{1}{2\pi i\lambda}\, Q(\tfrac{1}{\lambda})\, d\lambda.$$

Then the functional calculus gives

$$\begin{aligned} g(P(a)) &= g(\textstyle\int_\Gamma \exp(\lambda a)\, d\mu_P(\lambda)) \\ &= \textstyle\int_\Gamma g(\exp(\lambda a))\, d\mu_P(\lambda) \\ &= \textstyle\int_\Gamma \phi(\lambda)\, d\mu_P(\lambda) = \Phi_\phi(P(u)). \end{aligned}$$

Suppose now that G is analytic on a neighbourhood D of K. We may choose a compact convex neighbourhood U of K with $U \subset D$. Then G can be approximated uniformly on U by polynomials, and therefore the continuity of the functional calculus (Hille-Phillips [164]) gives

$$g(G(a)) = \Phi_\phi(G(u)).$$

Since this holds for all $g \in A'$ with $\|g\| = 1$, we have

$$\|G(a)\| \leq \|\phi\| . \|G(u)\| \leq \|G(u)\|,$$

and (i) is proved. Also $\phi(0) = g(1)$, which completes the proof of (ii).

Corollary 10. (Bollobás [116], Browder [125], Crabb [132].) Let $K = \{z : |z| \leq 1\}$, and let n be a positive integer. Then for elements a of complex unital Banach algebras A,

(i) $\sup\{ \|a^n\| : v(a) \le 1 \} = n!\,(e/n)^n$;

(ii) for all a with $v(a) \le 1$,

$$V(A, a^n) \subset V(Ea(K), u^n) = \{z : |z| \le c_n\},$$

where

$$c_n = \sup\{ |\phi^{(n)}(0)| : \phi \in D(K),\ \phi(0) = \|\phi\| = 1 \}.$$

Proof. By NRI Theorem 4. 8, we know that $v(a) \le 1$ implies $\|a^n\| \le n!\,(e/n)^n$. The condition $v(a) \le 1$ is equivalent to the condition that (A, a) be subordinate to K. Also $\omega(\lambda) = e^{|\lambda|}$, and $D(K)$ contains all polynomials. Let Γ be a circle with centre at 0 and radius > 1. We have

$$z^n = \frac{n!}{2\pi i} \int_\Gamma e^{z\lambda} \frac{1}{\lambda^{n+1}}\, d\lambda \qquad (z \in K),$$

and hence $u^n = f_\mu$ with $d\mu(\lambda) = \frac{n!}{2\pi i} \frac{1}{\lambda^{n+1}}\, d\lambda$ on Γ. Therefore, for all $\phi \in D(K)$, we have

$$\begin{aligned}\Phi_\phi(u^n) &= \int \phi(\lambda)\, d\mu(\lambda)\\ &= \frac{n!}{2\pi i} \int_\Gamma \phi(\lambda) \frac{1}{\lambda^{n+1}}\, d\lambda\\ &= \phi^{(n)}(0).\end{aligned}$$

In particular, consider the function ϕ given by

$$\phi(\lambda) = (e\lambda/n)^n \qquad (\lambda \in \underset{\sim}{C}).$$

We have

$$|\phi(\lambda)|/\omega(\lambda) = \Big(\frac{|\lambda|}{n}\, e^{1-\frac{|\lambda|}{n}}\Big)^n \le 1,$$

so that $\|\phi\| \le 1$. Also

$$\Phi_\phi(u^n) = \phi^{(n)}(0) = n!\,\Big(\frac{e}{n}\Big)^n,$$

which gives $\|u^n\| \ge n!\,(\frac{e}{n})^n$, and completes the proof of (i).

Since $\Phi_\phi(u^n) = \phi^{(n)}(0)$ for all $\phi \in D(K)$, we have

$$V(Ea(K),\ u^n) = \{\phi^{(n)}(0) : \phi \in D(K),\ \phi(0) = \|\phi\| = 1\}.$$

Given $\phi \in D(K)$ and $\theta \in \underset{\sim}{R}$, consider ψ defined by

$$\psi(\lambda) = \phi(\lambda e^{i\theta}) \qquad (\lambda \in \underset{\sim}{C}).$$

Then $\psi \in D(K)$, $\psi(0) = \phi(0)$, and $\|\psi\| = \|\phi\|$. Since $\psi^{(n)}(0)=e^{ni\theta}\phi^{(n)}(0)$, this shows that $V(Ea(K),\ u^n)$ is a disc with its centre at 0, and completes the proof.

Remarks. (1) In the proof of this Corollary we have appealed to NRI Theorem 4. 8. Alternatively we note that by Theorem 9 (i) it is sufficient to prove that $\|u^n\| \le n!\ (\frac{e}{n})^n$, or equivalently that for all entire functions ϕ with $|\phi(\lambda)| \le e^{|\lambda|}$ $(\lambda \in \underset{\sim}{C})$, we have

$$|\phi^{(n)}(0)| \le n!\ (e/n)^n.$$

This is an elementary application of the Cauchy estimates as shown in the proof of Lemma 4, Case 1.

(2) The proof of the Corollary given by Browder [125] amounts to a consideration of the differentiation operator on $D(K)$; in fact if

$$T_u f = uf \qquad (f \in Ea(K)),$$

Then T_u^* is the differentiation operator on $D(K)$. The proof of the Corollary given by Crabb [132] shows that each $f \in Ea(K)$ has a discrete representing measure, i. e. we have

$$f(z) = \Sigma c_n e^{\lambda_n z} \qquad (z \in K),$$

where $\Sigma |c_n| \omega(\lambda_n) < \infty$.

(3) In the proof of the Corollary we established that

$$\Phi_\phi(u^n) = \phi^{(n)}(0).$$

Suppose that $\sum_{n=0}^{\infty} c_n u^n$ converges to $f \in Ea(K)$ in $\|\cdot\|$ (and so also in $\|\cdot\|_\infty$). Then

$$\Phi_\phi(f) = \sum_{n=0}^{\infty} c_n \phi^{(n)}(0) = \sum_{n=0}^{\infty} \frac{1}{n!} f^{(n)}(0)\phi^{(n)}(0).$$

(4) Consider the general case with $0 \in \text{int } K$. Given G analytic on a neighbourhood of K and any representation of G by

$$G(z) = \int e^{\lambda z} \, d\mu(\lambda),$$

we have

$$G(a) = \int \exp(\lambda a) \, d\mu(\lambda).$$

To see this, let $0 < t < 1$, $\Gamma = t^{-1}\partial K$. We have

$$G(ta) = \frac{1}{2\pi i} \int_\Gamma G(tz)(z - a)^{-1} dz$$

$$= \int \frac{1}{2\pi i} \int_\Gamma e^{t\lambda z}(z - a)^{-1} dz \, d\mu(\lambda)$$

$$= \int \exp(t\lambda a) \, d\mu(\lambda).$$

The result follows by continuity. This simple technique does not apply when $K \subset \underset{\sim}{R}$; for the Hermitian case alternative techniques are developed in § 26.

(5) Now let K be arbitrary. Following a suggestion of S. Kaijser, the authors show in [134] that $Ea(K)$ is itself a dual space and its predual may be identified as

$$D_0(K) = \{\phi \in D(K) : \lim_{|\lambda| \to 0} |\phi(\lambda)|/\omega(\lambda) = 0\}.$$

25. OTHER EXTREMAL ALGEBRAS

In the previous section we constructed the extremal algebra subject to the condition $V(a) \subset K$. For the special case $K \subset \underset{\sim}{R}$ this corresponds to the study of extremal properties of Hermitian elements and this topic is taken up in the following sections. A further special case was considered in §23 where we studied some of the extremal properties of numerical ranges of idempotent elements. In this section we consider one further special case, the case of normal elements. We also describe the extremal algebras subject to the global condition $V(a) = \text{co } Sp(a)$ $(a \in A)$.

Let A be a complex unital Banach algebra. We recall that $a \in A$ is normal if $a = h + ik$ where $h, k \in H(A)$, $hk = kh$. For a normal element a we have

$$V(a) = \text{co } Sp(a)$$

by NRI Theorem 5.14, and hence $v(a) = \rho(a)$. The natural analogue of §24 would be to study the extremal algebras subject to the conditions that a be normal and $V(a) \subset K$, where K is a compact convex subset of $\mathbb{C}$. For simplicity of exposition, we shall confine our attention to the problem of finding the maximum value of $\|a\|$ subject to the above condition with K the closed unit disc. For a fuller study of normal elements we refer the reader to [134].

Lemma 1. Let a be normal with $\rho(a) = 1$. Then $\|a\| \leq 2$.

Proof. Let $a = h + ik$ with $h, k \in H(A)$, $hk = kh$. Since h, k have real spectrum it follows that $\rho(h) \leq 1$, $\rho(k) \leq 1$. Using Theorem 26.2 (which does not depend on the present section) we deduce that

$$\|a\| \leq \|h\| + \|k\| = \rho(h) + \rho(k) \leq 2.$$

We shall show that Lemma 1 is best possible. The construction of the extremal algebra appropriate to this problem follows the pattern of §24. To illustrate an alternative method of proof we adopt a slightly different approach from §24. Let

$$E = \{(s,t) : s, t \in \mathbb{R},\ s^2 + t^2 \leq 1\}.$$

We denote by h, k the functions defined on E by

$$h(s,t) = s, \qquad k(s,t) = t,$$

and by ω the function defined on $\mathbb{C}^2$ by

$$\omega(z,w) = \|\exp(zh + wk)\|_\infty \qquad (z, w \in \mathbb{C}),$$

where $\|\cdot\|_\infty$ is the supremum norm for functions on E. We have

$$\omega(z,w) = \exp\{[(\text{Re } z)^2 + (\text{Re } w)^2]^{\frac{1}{2}}\}.$$

We denote by $M^e(\underset{\sim}{C}^2)$ the set of all finite complex regular Borel measures μ on $\underset{\sim}{C}^2$ such that

$$\int e^{e(|z|+|w|)} d|\mu|(z,w) < \infty.$$

Given $\mu \in M^e(\underset{\sim}{C}^2)$ we define a function f_μ on E by

$$f_\mu(s,t) = \int e^{zs+wt} d\mu(z,w) \qquad ((s,t) \in E)$$

and we denote by B the set of all functions f_μ with $\mu \in M^e(\underset{\sim}{C}^2)$. Given $f \in B$ we define $\|f\|$ by

$$\|f\| = \inf\{\int \omega d|\mu| : \mu \in M^e(\underset{\sim}{C}^2),\ f_\mu = f\}.$$

Theorem 2. $(B, \|\cdot\|)$ _is a complex unital normed algebra of continuous complex valued functions on_ E, _and_

(i) $\|f\|_\infty \le \|f\| \qquad (f \in B)$,

(ii) $h, k \in B$,

(iii) _for_ $z, w \in \underset{\sim}{C}$, _the exponential of_ $zh + wk$ _is an element of_ B _which coincides with the function_ $\exp(zh + wk)$ _and satisfies_

$$\|\exp(zh + wk)\|_\infty = \|\exp(zh + wk)\|.$$

Proof. Apply the methods of Theorem 24. 2.

Corollary 3. _The element_ $h + ik$ _is normal in_ B.

Proof. From (iii) above, $\|\exp(ith)\| = \|\exp(itk)\| = 1$ for $t \in \underset{\sim}{R}$ and clearly B is commutative.

Lemma 4. _Let_ ϕ _be an entire function of two variables, let_ $|\phi| \le \omega$, _and let_ $\mu \in M^e(\underset{\sim}{C}^2)$. _If_ $f_\mu = 0$, _then_ $\int \phi\, d\mu = 0$.

Proof. Since

$$\int \sum_{m,n=0}^{\infty} \frac{s^m t^n}{m!\,n!} z^m w^n d\mu(z,w) = 0 \qquad ((s,t) \in E),$$

it follows by dominated convergence that

$$\int z^m w^n d\mu(z,w) = 0 \qquad (m, n = 0, 1, 2, \dots). \qquad (1)$$

Let $\sum_{m,n=0}^{\infty} c_{mn} z^m w^n$ be the power series representation of $\phi(z,w)$. Since $|\phi| \le \omega$ the Cauchy estimates give

$$|c_{mn}| \le \frac{e^r}{r^m} \frac{e^s}{s^n} \qquad (r, s > 0),$$

and so $|c_{mn}| \le (e/m)^m (e/n)^n$. It follows that

$$\sum_{m,n=0}^{\infty} |c_{mn} z^m w^n| \le e^{e(|z|+|w|)} \qquad (z, w \in \mathbb{C}),$$

and so (1) and dominated convergence give $\int \phi \, d\mu = 0$.

We need one more lemma before the main result.

Lemma 5. Let ϕ be an entire function of two variables such that

(i) $|\phi(z,w)| \le e^{(|z|+|w|)}$ $\quad (z, w \in \mathbb{C})$,

(ii) $|\phi(z,w)| \le 1$ $\quad (z, w \in i\mathbb{R})$,

(iii) $|\phi(z \cos\theta - w \sin\theta, z \sin\theta + w\cos\theta)| = |\phi(z,w)|$ $(z, w \in \mathbb{C}, \theta \in \mathbb{R})$.

Then $|\phi(z,w)| \le \omega(z,w)$ $(z, w \in \mathbb{C})$.

Proof. Given $w \in i\mathbb{R}$ we have

$$|\phi(z,w)| \le e^{|w|} e^{|z|} \qquad (z \in \mathbb{C})$$

$$|\phi(z,w)| \le 1 \qquad (z \in i\mathbb{R}),$$

and so the Phragmen-Lindelöf principle gives

$$|\phi(z,w)| \le e^{|\mathrm{Re}\, z|} \qquad (z \in \mathbb{C}).$$

Given arbitrary $z, w \in \mathbb{C}$, choose $\theta \in \mathbb{R}$ such that $z \sin\theta + w \cos\theta \in i\mathbb{R}$. Then

$$
\begin{aligned}
|\phi(z,w)| &= |\phi(z\cos\theta - w\sin\theta,\ z\sin\theta + w\cos\theta)| \\
&\le e^{|\mathrm{Re}(z\cos\theta - w\sin\theta)|} \\
&\le \omega(z,w).
\end{aligned}
$$

Theorem 6. <u>The normal element</u> $h + ik$ <u>of</u> B <u>satisfies</u>

$$\rho(h + ik) = 1, \qquad \|h + ik\| = 2.$$

Proof. For $z \in \underset{\sim}{C}$, Theorem 2 (iii) gives

$$\|\exp z(h + ik)\| = \|\exp(zh + izk)\|_\infty = e^{|z|},$$

and therefore $\rho(h + ik) \le 1$. The spectrum of $h + ik$ in the completion of B contains the closed unit disc by Theorem 2 (i), and hence $\rho(h+ik)=1$.

Let ϕ be defined on $\underset{\sim}{C}^2$ by

$$\phi(z,w) = (z - iw)\sum_{n=0}^{\infty} \frac{(z^2+w^2)^n}{(2n+1)!}$$

It is easily verified that ϕ satisfies the conditions of Lemma 5 and hence $|\phi| \le \omega$. By Lemma 4 we may define a functional Φ on B by

$$\Phi(f) = \int \phi\, d\mu \qquad (\mu \in M^e(\underset{\sim}{C}^2),\ f_\mu = f),$$

and it is clear that

$$|\Phi(f)| \le \|f\| \qquad (f \in B).$$

Since (with Γ the unit circle)

$$(h + ik)(s,t) = \left(\frac{1}{2\pi i}\right)^2 \int_\Gamma\int_\Gamma e^{zs+wt}\left(\frac{dz}{z^2}\frac{dw}{w} + i\frac{dz}{z}\frac{dw}{w^2}\right),$$

we have

$$
\begin{aligned}
\Phi(h + ik) &= \left(\frac{1}{2\pi i}\right)^2 \int_\Gamma\int_\Gamma \phi(z,w)\left(\frac{dz}{z^2}\frac{dw}{w} + i\frac{dz}{z}\frac{dw}{w^2}\right) \\
&= 2, \\
\|h + ik\| &\ge 2.
\end{aligned}
$$

Lemma 1 gives $\|h + ik\| \le 2$. The proof is complete.

Remarks. (1) It can be deduced from Theorem 26.9 that there is exactly one continuous functional Φ on B such that $\|\Phi\| = 1$, $\Phi(h + ik) = 2$.

(2) It is shown in [134] that

$$\|(h + ik)^n\| = 2^n \sup\{|g^{(n)}(0)| : g \in G_n\}$$

where

$$G_n = \{g \in D([-1,1]) : \|g\| = 1,\ g(0) = g'(0) = \ldots = g^{(n-1)}(0) = 0\}.$$

Taking g to be the Bessel function J_n, we have

$$\|(h + ik)^n\| \geq M\, n^{1/3} \qquad (n = 1, 2, \ldots)$$

for some positive constant M. It follows that the power inequality fails for the normal element $h + ik$ of B.

Finally we consider the global condition

$$V(a) = \operatorname{co} Sp(a) \qquad (a \in A) \tag{2}$$

where A is a complex unital Banach algebra. We have seen in Lemma 23.4 that condition (2) is equivalent to the condition

$$\|\exp(a)\| = \rho(\exp(a)) \qquad (a \in A).$$

By NRI Theorem 4.7, A is then a function algebra under an equivalent norm. In fact, Theorem 23.5 gives the estimate

$$\rho(a) \leq \|a\| \leq \tfrac{1}{2}e\, \rho(a) \qquad (a \in A),$$

and we show below that the constant $\frac{1}{2}e$ is best possible in general.

We shall confine our attention to singly generated Banach algebras, so that the spectrum of the generator has connected complement. Let T be any compact subset of $\underset{\sim}{C}$ with connected complement and let A(T) be the algebra of continuous functions on T that are analytic on int T. For $f \in A(T)$ we define $\|f\|$ by

$$\|f\| = \inf\{\Sigma|c_n|\rho(g_n) : g_n \in \exp A(T),\ f = \Sigma c_n g_n\}$$

where $f = \Sigma c_n g_n$ means that the series converges to f uniformly on T.

Theorem 7. $(A(T), \|\cdot\|)$ is a complex unital Banach algebra with $\|\exp(f)\| = \rho(\exp(f))$ $(f \in A(T))$.

Proof. Let $f \in A(T)$ with $\rho(f) < 1$. Since

$$f = \tfrac{1}{2}(1 + f) - \tfrac{1}{2}(1 - f)$$

and $\exp A(T) = \{g \in A(T) : 0 \notin g(T)\}$, it follows that each $f \in A(T)$ has a representation of the required form. Now argue as in the proof of Theorem 24.2.

Theorem 8. Let p be a complex Banach algebra norm on $A(T)$ such that

$$p(\exp(f)) = \rho(\exp(f)) \qquad (f \in A(T)).$$

Then $\rho(f) \leq p(f) \leq \|f\|$ $(f \in A(T))$.

Proof. Let $f \in A(T)$ and let $f = \Sigma c_n g_n$ where $g_n \in \exp A(T)$, $\Sigma |c_n| \rho(g_n) < \infty$. Then

$$p(f) \leq \Sigma |c_n| \, p(g_n) = \Sigma |c_n| \, \rho(g_n)$$

and so $p(f) \leq \|f\|$, as required.

Given $f \in A(T)$ we have $V(f) = \text{co}\,\text{Sp}(f) = \text{co}\, f(T)$. For the case $\text{int}\, T = \emptyset$ we have $A(T) = C(T)$ and

$$H(A(T)) = C_{\underset{\sim}{R}}(T).$$

The Vidav-Palmer theorem (NRI Theorem 6.9) then gives

$$\|f\| = \rho(f) \qquad (f \in A(T)).$$

More generally, given any compact Hausdorff space E, the spectral norm is the only complex Banach algebra norm on $C(E)$ such that condition (2) holds with $A = C(E)$. We shall exhibit a class of simply connected compact sets T for which the extremal norm on $A(T)$ differs from the spectral norm and the estimate in Theorem 23.4 is best possible. As before we write $u(z) = z$ $(z \in T)$.

Lemma 9. *When T is the closed unit disc,* $\|u\| = \frac{1}{2}e$.

Proof. Let $\Phi(f) = f'(0)$ $(f \in A(T))$. Clearly $\Phi \in A(T)'$ and $\Phi(u) = 1$. It is now sufficient to show that

$$|\Phi(f)| \leq 2e^{-1}\|f\| \qquad (f \in A(T)).$$

By the definition of the extremal norm $\|\cdot\|$, it is sufficient to show that

$$|\exp(f(0))\, f'(0)| = |\Phi(\exp(f))| \leq 2e^{-1}\rho(\exp f) \qquad (f \in A(T)),$$

or equivalently

$$|g'(0)| \leq 2e^{-1}\rho(\exp g) \qquad (g \in A(T) \quad g(0) = 0). \tag{3}$$

Suppose that $\operatorname{Re} g \leq 1$ and let $h = g/(2 - g)$. Then $h \in A(T)$, $h(T) \subset T$, $h(0) = 0$. By Schwarz's Lemma, $|h'(0)| \leq 1$ and so $|g'(0)| = |2h'(0)| \leq 2$. Given arbitrary $g \in A(T)$ with $g(0) = 0$, let $\delta = \sup \operatorname{Re} g(T)$. Note that $\delta > 0$ since $0 \in \operatorname{int} g(T)$. Let $k = e^{1-\delta}g$, so that $k(0) = 0$, $\operatorname{Re} k \leq 1$. Then $|g'(0)| \leq 2e^{-1}e^{\delta}$, which is the required inequality (3).

Corollary 10. *Let T be the image of the closed unit disc under a homeomorphism* σ *which is analytic on the interior of the disc, and let* $\tau = \sigma^{-1}$. *Then* $\tau \in A(T)$, $\rho(\tau) = 1$, *and* $\|\tau\| = \frac{1}{2}e$.

Proof. It is easily verified that σ induces an isometric isomorphism $f \to f \circ \sigma$ between $A(T)$ for the given T and the disc algebra. The result follows by Lemma 9.

26. PROPERTIES OF HERMITIAN ELEMENTS

Let A be a complex unital Banach algebra and let $h \in H(A)$, i.e. let h be Hermitian. Vidav's lemma (NRI Theorem 5.10) gives $\rho(h) = v(h)$, and it was announced in Volume I of these notes (page 54) that Sinclair [62] had further proved that $\rho(h) = \|h\|$. This result had also been proved independently by Browder [125], [126], and an elementary proof was subsequently given by Bonsall and Crabb [122]. This latter proof is given below in modified form. Sinclair [62] in fact proved the stronger result

$$\rho(\alpha + \beta h) = \|\alpha + \beta h\| \qquad (\alpha, \beta \in \mathbb{C}) \tag{1}$$

using a version of the Phragmen-Lindelöf theorem that was established by Duffin and Schaeffer [145]. Bollobás (lecture) has also established (1) using the subordination theory of Levin [174].

We shall establish (1) by obtaining a generalized Fourier expansion of $\cos\theta + (i\sin\theta)h$ when $\|h\| = 1$. This expansion was obtained by Crabb (see [134]), Bonsall [120] and others, but we shall use the method of Browder [126]. As an application of the Fourier expansion we derive the even-odd theorem (Theorem 9) which describes the support functionals at $\cos\theta + (i\sin\theta)h$. The even-odd theorem is due to Crabb (see [134]), Bollobás [118] and (in a special case) Sinclair [202].

Let $\{c_r\}$ be the sequence of non-negative real numbers for which

$$\arcsin t = \sum_{r=1}^{\infty} c_r t^r \qquad (|t| \leq 1),$$

and let

$$F_n(z) = \sum_{r=1}^{n} c_r (\sin z)^r \qquad (z \in \mathbb{C}).$$

Lemma 1. Let K be a compact subset of $(-\frac{1}{2}\pi, \frac{1}{2}\pi)$. Then there is an open subset U of $\mathbb{C}$ such that $K \subset U$ and

$$F_n(z) \to z \qquad \text{(uniformly on } U).$$

Proof. There is an open set U of $\mathbb{C}$ with $K \subset U$ and $|\sin z| \leq 1$ $(z \in U)$. By the Weierstrass M-test, $\sum_{r=1}^{\infty} c_r (\sin z)^r$ converges uniformly on U and so the limit function is analytic on U. We have $\lim_{n\to\infty} F_n(t) = t$ $(t \in K)$ by real analysis and the result follows.

Theorem 2. Let $h \in H(A)$. Then $\rho(h) = \|h\|$.

Proof. Let $\rho(h) < \frac{1}{2}\pi$. Then $\mathrm{Sp}(h) \subset (-\frac{1}{2}\pi, \frac{1}{2}\pi)$, and it follows from Lemma 1 and the functional calculus that

$$h = \lim_{n\to\infty} F_n(h) = \sum_{n=1}^{\infty} c_n (\sin h)^r.$$

Since $h \in H(A)$, we have $\|\sin h\| \le \frac{1}{2}\|\exp(ih)\| + \frac{1}{2}\|\exp(-ih)\| = 1$, and therefore

$$\|h\| \le \sum_{n=1}^{\infty} c_n = \frac{1}{2}\pi.$$

The result follows by the positive homogeneity of the spectral radius and the norm.

Lemma 3. Let F be an entire function such that

$$|F(z)| \le \exp(|\operatorname{Im} z|) \qquad (z \in \mathbb{C}),$$

and let $\theta \in \mathbb{R}\backslash\pi\mathbb{Z}$. Then

$$F(0)\cos\theta + F'(0)\sin\theta = \sum_{n\in\mathbb{Z}} \gamma_n F(n\pi + \theta),$$

where

$$\gamma_n = \frac{(-1)^n}{(n\pi+\theta)^2}\sin^2\theta \qquad (n \in \mathbb{Z}). \tag{2}$$

Proof. Let Γ_n be the boundary of the square with vertices $\pm a_n \pm ia_n$, where $a_n = |\theta| + (n + \frac{1}{2})\pi$. A straightforward estimate gives

$$\lim_{n\to\infty} \int_{\Gamma_n} \frac{F(z)dz}{z^2\sin(\theta - z)} = 0.$$

Now apply Cauchy's residue theorem.

Theorem 4. Let $h \in H(A)$ with $\|h\| \le 1$, let $\theta \in \mathbb{R}\backslash\pi\mathbb{Z}$, and let $\{\gamma_n\}$ be as in (2). Then

$$\cos\theta + (i\sin\theta)h = \sum_{n\in\mathbb{Z}} \gamma_n \exp((n\pi + \theta)ih). \tag{3}$$

Proof. Let $g \in A'$ with $\|g\| \leq 1$, and let

$$F(z) = g(\exp(izh)) \qquad (z \in \mathbb{C}).$$

Then F is entire, and for $z \in \mathbb{C}$

$$\begin{aligned} |F(z)| &\leq \|g\| \; \|\exp(i \operatorname{Re} z\, h)\| \; \|\exp(-\operatorname{Im} z\, h)\| \\ &\leq \exp(|\operatorname{Im} z|\, v(h)) \\ &\leq \exp(|\operatorname{Im} z|). \end{aligned}$$

Lemma 3 gives

$$\begin{aligned} g(\cos\theta + (i\sin\theta)h) &= \sum_{n\in\mathbb{Z}} \gamma_n g(\exp((n\pi + \theta)ih)) \\ &= g\Big(\sum_{n\in\mathbb{Z}} \gamma_n \exp((n\pi+\theta)ih)\Big), \end{aligned}$$

and the Hahn-Banach theorem completes the proof.

Corollary 5. Let $h \in H(A)$ with $\|h\| \leq 1$. Then

$$h = \sum_{n\in\mathbb{Z}} 4\pi^{-2}(2n+1)^{-2} \exp\{(n+\tfrac{1}{2})\pi i(h-1)\}.$$

Proof. Let $\theta = \frac{1}{2}\pi$ in (3).

Remark. Note that $\sum_{n\in\mathbb{Z}} 4\pi^{-2}(2n+1)^{-2} = 1$.

Corollary 6. Let $h \in H(A)$. Then

$$\rho(\alpha + \beta h) = \|\alpha + \beta h\| \qquad (\alpha, \beta \in \mathbb{C}).$$

Proof. Let $\|h\| = 1$. Then by (3)

$$\|\cos\theta + (i\sin\theta)h\| \leq \sum_{n\in\mathbb{Z}} |\gamma_n| = 1.$$

We have $\rho(h) = 1$, by Theorem 2, and so $\rho(\cos\theta + (i\sin\theta)h) \geq 1$. This gives $\rho(\cos\theta + (i\sin\theta)h) = \|\cos\theta + (i\sin\theta)h\|$ and the result follows easily.

As an application of Corollary 6 we deduce the result of Bollobás [118] that a Hermitian isometry is its own inverse.

Theorem 7. Let $h \in H(A)$ be invertible with $\|h\| = \|h^{-1}\| = 1$. Then $h = h^{-1}$.

Proof. For all $t \in \mathbb{R}$ we have

$$\|1 + ith^{-1}\| = \|h + it\| = (1 + t^2)^{\frac{1}{2}},$$

and so $h^{-1} \in H(A)$ by NRI Lemma 5.2. Theorem 2 gives

$$\|h - h^{-1}\| = \rho(h - h^{-1}).$$

Since $Sp(h) \subset \{1, -1\}$, it follows that $h = h^{-1}$.

Lemma 8. Let F be an entire function such that $F(n) = 0$ $(n \in \mathbb{Z})$ and

$$|F(z)| \le M \exp(|\mathrm{Im}\, \pi z|) \qquad (z \in \mathbb{C}).$$

Then $F(z) = F(\frac{1}{2}) \sin \pi z$ $(z \in \mathbb{C})$.

Proof. Let $G(z) = F(z)/\sin \pi z$ $(z \in \mathbb{C})$. Then G is entire and bounded, and so constant by Liouville's theorem.

Theorem 9. Let $h \in H(A)$, $\|h\| = 1$, let $\theta \in \mathbb{R} \backslash \pi\mathbb{Z}$ and let $f \in A'$ be such that $f(\cos\theta + (i \sin\theta)h) = 1 = \|f\|$. Then there exists $c \in [-1, 1]$ such that

$$f(h^n) = \begin{cases} -i \sin\theta + c\cos\theta & \text{if } n = 1, 3, 5, \ldots \\ \cos\theta - ic\sin\theta & \text{if } n = 0, 2, 4, \ldots . \end{cases}$$

Proof. By Theorem 4, we have

$$1 = \sum_{n \in \mathbb{Z}} \gamma_n f(\exp((n\pi + \theta)ih)),$$

and, since $\sum_{n \in \mathbb{Z}} (-1)^n \gamma_n = 1$ and

$$|f(\exp((z\pi + \theta)ih)| \le 1 \qquad (z \in \mathbb{R}), \tag{4}$$

it follows that

$$f(\exp((n\pi + \theta)ih)) = (-1)^n \qquad (n \in \underset{\sim}{Z}).$$

Let

$$F(z) = f(\exp((\pi z + \theta)ih)) - \cos \pi z \qquad (z \in \underset{\sim}{C}).$$

Then F is entire, $F(n) = 0 \ (n \in \underset{\sim}{Z})$, and

$$|F(z)| \leq M \exp(|\mathrm{Im}\ \pi z|) \qquad (z \in \underset{\sim}{C}),$$

and so $F(z) = F(\frac{1}{2}) \sin \pi z$, by Lemma 8. Therefore

$$f(\exp((\pi z + \theta)ih)) = \cos \pi z + F(\tfrac{1}{2}) \sin \pi z \qquad (z \in \underset{\sim}{C}).$$

It follows from (4) that $F(\frac{1}{2}) = ic$ for some $c \in [-1, 1]$, and so

$$f(\exp(izh)) = \cos(z - \theta) + ic \sin(z - \theta).$$

The proof is completed by comparing coefficients of z^n in this last equation.

Corollary 10. Let $h \in H(A)$, $f \in A'$ be such that

$$f(h) = f(1) = 1 = \|f\| = \|h\|.$$

Then f is multiplicative on the subalgebra of A generated by 1 and h.

Proof. Let $g(x) = f(ix) \ (x \in A)$ and apply Theorem 9 to g with $\theta = -\frac{1}{2}\pi$ to give $f(h^n) = 1$. The result follows by simple computation.

Let $\mathscr{F}$ be the set of entire functions F such that

$$|F(z)| \leq \exp(|\mathrm{Im}\ z|) \qquad (z \in \underset{\sim}{C}).$$

It was pointed out by Bollobás [117] and Browder [125] that the result

$$v(h) = \|h\| \qquad (h \in H(A))$$

is equivalent to Bernstein's theorem that

$$F \in \mathscr{F} \Rightarrow F' \in \mathscr{F}.$$

Thus the properties of Hermitian elements are reflected by the properties of the family $\mathscr{F}$ of entire functions. Using such properties and the methods of this section, Bollobás [117], Browder [125] and Crabb (see [134]) have obtained norm properties of other polynomials of Hermitian elements. For example, Crabb has shown that given $h \in H(A)$ with $\|h\| = 1$ we have

$$\|h^2 - t\| = 1 - t = \rho(h^2 - t) \qquad (0 \leq t \leq \tfrac{1}{3}).$$

The next section studies the global properties of polynomials in h as reflected in the closed subalgebra generated by h.

27. BANACH ALGEBRAS GENERATED BY A HERMITIAN ELEMENT

Let A be a unital Banach algebra over $\underset{\sim}{C}$, and suppose that A is generated by a Hermitian element h of A, i.e. A is the least closed subalgebra of A containing 1 and h.

Sinclair [204] has proved that A is semi-simple if $Sp(h)$ is countable, but is not semi-simple in general. We give here an account of some of his results. Let J denote the Jacobson radical of A, and Φ_A the Gelfand carrier space of all non-zero multiplicative linear functionals on A.

Lemma 1. $J = \cap\{[(h - \lambda)A]^- : \lambda \in Sp(h)\}$.

Proof. Let $j \in J$. Then there exists a sequence $\{p_n\}$ of non-constant polynomials p_n such that $j = \lim_{n \to \infty} p_n(h)$. Given $\lambda \in Sp(h)$, there exists $\phi \in \Phi_A$ with $\phi(h) = \lambda$. Then

$$0 = \phi(j) = \lim_{n \to \infty} \phi(p_n(h)) = \lim_{n \to \infty} p_n(\lambda).$$

We have $p_n(h) - p_n(\lambda) = (h - \lambda)b_n$ for some $b_n \in A$, and so $j = \lim_{n \to \infty} p_n(h) - p_n(\lambda) \in [(h - \lambda)A]^-$. We have proved that

$$J \subset \cap\{[(h - \lambda)A]^- : \lambda \in Sp(h)\}.$$

Given $\phi \in \Phi_A$, we have $\phi(h) = \lambda \in Sp(h)$, and

$$[(h - \lambda)A]^- \subset Ker(\phi).$$

Therefore

$$\cap\{[(h - \lambda)A]^- : \lambda \in Sp(h)\} \subset \cap\{Ker(\phi) : \phi \in \Phi_A\} = J.$$

Lemma 2. If $\lambda \in Sp(h)$, $j \in J$, and $(h - \lambda)j = 0$, then $j = 0$.

Proof. Let $\lambda \in Sp(h)$, and let $T \in B(A)$ be defined by $Ta = (h - \lambda)a$ $(a \in A)$. Then T is a Hermitian operator, and so all points of the spectrum are in the boundary of its numerical range, and orthogonality gives

$$Ker(T) \cap (TA)^- = \{0\}.$$

But by Lemma 1, $J \subset (TA)^-$, and so $Ker(T) \cap J = \{0\}$.

Theorem 3. Let $Sp(h)$ be countable. Then A is semi-simple.

Proof. Suppose that $J \neq \{0\}$. Then J is a non-zero Banach space. Given $a \in A$, define $T_a \in B(J)$ by

$$T_a j = aj \quad (j \in J).$$

Then $a \to T_a$ is a homomorphism of A into $B(J)$. Therefore

$$Sp(T_a) \subset Sp(a).$$

Also, given $(x, f) \in \Pi(J)$, the functional g defined on A by

$$g(a) = f(ax) \qquad (a \in A)$$

belongs to $D(1)$, and so

$$V(T_a) \subset V(A, a) \qquad (a \in A).$$

It follows in particular that $Sp(T_h)$ is countable and that T_h is a Hermitian operator.

Since a compact space is second category in itself, at least one point of $Sp(T_h)$ constitutes a relatively open subset of $Sp(T_h)$. Therefore $Sp(T_h)$ has an isolated point, μ say. Therefore there exists a non-zero projection $E \in B(J)$ such that if U denotes the restriction of T_h to EJ, we have

$$Sp(U) = \{\mu\}.$$

But U is a Hermitian operator on EJ, and so, by the Sinclair theorem

$$\|\mu I - U\| = \rho(\mu I - U) = 0.$$

Thus $(\mu I - T_h)EJ = \{0\}$, from which

$$(\mu - h)x = 0 \qquad (x \in EJ).$$

But then, by Lemma 2, $EJ = \{0\}$, $E = 0$, a contradiction.

In [204] Corollary 3.4, Sinclair gives a class of examples of Hermitian operators that generate Banach algebras with non-trivial radicals. The proof uses quite difficult ideas from harmonic analysis; the theorem below indicates how the problem is reduced to the detailed study of the following group algebra. Recall that $l^1(\underset{\sim}{R})$ is the L^1 group algebra of the discrete additive group of reals. We also note from Theorem 24.2 that $Ea([-1,1])$ is generated by the Hermitian element u, where $u(t) = t \quad (-1 \le t \le 1)$.

Theorem 4. $Ea([-1,1])$ <u>is isometrically isomorphic to a quotient of</u> $l^1(\underset{\sim}{R})$.

Proof. Since $\|\exp(i\alpha u)\| = 1 \quad (\alpha \in \underset{\sim}{R})$ we may define θ from $l^1(\underset{\sim}{R})$ to $Ea([-1,1])$ by

$$\theta(\lambda) = \Sigma\{\lambda(\alpha)\exp(i\alpha u) : \alpha \in \underset{\sim}{R}\}.$$

It is clear that θ is a homomorphism with $\|\theta\| \le 1$, and it is onto by Corollary 26.5. The kernel K of θ is the closed ideal defined by

$$K = \{\lambda \in l^1(\underset{\sim}{R}) : \sum_{\alpha \in \underset{\sim}{R}} \lambda(\alpha)\, e^{i\alpha t} = 0 \qquad (-1 \le t \le 1)\},$$

and $Ea([-1,1])$ is isomorphic to $l^1(\mathbb{R})/K$. Let A be the algebra $Ea([-1,1])$ with the norm $\|\cdot\|'$ induced from the quotient norm on $l^1(\mathbb{R})/K$. Then A is a Banach algebra and

$$\|f\| \leq \|f\|' \qquad (f \in Ea([-1,1])),$$

since $\|\theta\| \leq 1$. By Banach's isomorphism theorem, the norms $\|\cdot\|'$ and $\|\cdot\|$ are equivalent, and therefore A is also generated by u. We have $\|\exp(i\alpha u)\|' \leq 1$ $(\alpha \in \mathbb{R})$, and Corollary 26.5 gives $\|u\|' \leq 1$. Therefore $V(A,u) \subset [-1,1]$, and Theorem 24.10 gives

$$\|p(u)\|' \leq \|p(u)\| \qquad (p \text{ a polynomial}),$$

and hence $\|f\|' \leq \|f\|$ $(f \in Ea([-1,1]))$, since u generates both $Ea([-1,1])$ and A. The proof is complete.

Sinclair [204] uses harmonic analysis to show that $Ea([-1,1])$ is a semi-simple regular Banach algebra satisfying a strong Ditkin condition and that spectral synthesis fails in $Ea([-1,1])$. Suitable quotient algebras of $Ea([-1,1])$ contain Hermitian elements which generate non-semi-simple Banach algebras. Sinclair also gives examples of Hermitian elements h such that the norm on A is equivalent to the spectral norm. This is not the case for the extremal algebra $Ea([-1,1])$.

Theorem 5. For the algebra $Ea([-1,1])$ we have

$$\|\exp(itu^2)\| \geq 2t^{\frac{1}{2}}/M \qquad (t > 0),$$

where $M = \sup_{a,b\in\mathbb{R}} \left| \int_a^b e^{-ix^2}\,dx \right|$.

Proof. Given $t > 0$, let ϕ be defined by

$$\phi(z) = \int_{-1}^{1} e^{zx} e^{-itx^2}\,dx \qquad (z \in \mathbb{C}).$$

Then $\phi \in D([-1,1])$, and $\Phi_\phi(\exp(itu^2)) = 2$. Using the Phragmen-Lindelöf theorem, we obtain

$$\|\phi\| = \sup_{s\in\mathbb{R}} |\phi(s)|$$

$$= \sup_{s \in \mathbb{R}} t^{-\frac{1}{2}} \left| \int_{-t^{\frac{1}{2}} - \frac{1}{2} s t^{-\frac{1}{2}}}^{t^{\frac{1}{2}} - \frac{1}{2} s t^{-\frac{1}{2}}} e^{-ix^2}\, dx \right|$$

$$\leq M t^{-\frac{1}{2}}.$$

The result follows by Theorem 24.5.

Given a Hermitian operator T on a Banach space X, let $R(T)$ be the weak operator closed subalgebra of $B(X)$ generated by T. Sinclair [204] shows that $R(T)$ is semi-simple if $Sp(T)$ is countable. Dowson [143] gives an example of a Hermitian operator T on a separable reflexive Banach space X such that the second commutant of $R(T)$ is strictly larger than $R(T)$.

28. COMPACT HERMITIAN OPERATORS

For a compact self-adjoint operator T on a Hilbert space there is the classical spectral decomposition given by

$$T = \Sigma \lambda_n P_n, \tag{1}$$

where $\{\lambda_n\}$ is the set of non-zero eigenvalues of T and P_n is the spectral projection corresponding to λ_n. In this section we investigate the extent to which equation (1) holds when T is a compact Hermitian operator on an arbitrary complex Banach space. The first theorem is an easy consequence of previous results on Hermitians. The other main result involves an exposition of work by Yu I. Lyubič [177] on the relation between Hermitian operators and uniformly almost periodic functions. The work of Lyubič goes beyond the context of compact Hermitian operators; we refer the reader to [177], [178], [179].

Throughout this section X denotes a complex Banach space.

Theorem 1. Let $T \in B(X)$ be compact and Hermitian, let $\{\lambda_n\}$ be the non-zero (distinct) eigenvalues of T arranged so that

$$|\lambda_{n+1}| \leq |\lambda_n| \qquad (n = 1, 2, \ldots),$$

<u>and let P_n be the spectral projection corresponding to λ_n. Then the following statements hold.</u>

(i) <u>Each eigenvalue of T has ascent</u> 1.

(ii) $|P_n| = 1$.

(iii) <u>If each P_n is Hermitian, then</u> $T = \Sigma\, \lambda_n P_n$.

(iv) <u>If</u> $\lim_{n\to\infty} n\lambda_n = 0$, <u>then</u> $T = \Sigma\, \lambda_n P_n$.

Proof. (i) Apply NRI Corollary 10.11.

(ii) By (i), P_nX is the null space and $(I - P_n)X$ is the range of $\lambda_n I - T$. Corollary 20.5 gives

$$\|P_n x + (I - P_n)y\| \geq \|P_n x\| \qquad (x, y \in X),$$

from which $\|x\| \geq \|P_n x\|$. Since P_n is a non-zero projection it follows that $|P_n| = 1$.

(iii) Consider the case in which there are infinitely many λ_n, the finite case being easier. Assume that each P_n is Hermitian, and let $S_n = \sum_{k=1}^{n} \lambda_k P_k$. Since each λ_k is real, $T - S_n$ is Hermitian and Theorem 26.2 gives

$$|T - S_n| = \rho(T - S_n) = |\lambda_{n+1}|.$$

But $\lim_{n\to\infty} \lambda_n = 0$, by the compactness of T.

(iv) With S_n as above and $R_n = \sum_{k=1}^{n} P_k$,

$$T - S_n = T_n(I - R_n),$$

where T_n is the restriction of T to $(I - R_n)X$. We have T_n Hermitian by Lemma 15.2, and so, by Theorem 26.2 again,

$$|T_n| = \rho(T_n) = |\lambda_{n+1}|.$$

But, by (ii), $|R_n| \leq n$ and so

$$|T - S_n| \leq (n + 1)|\lambda_{n+1}|.$$

Remarks. (1) Since

$$I - R_n = (I - P_n)(I - P_{n-1}) \cdots (I - P_1),$$

the proof used in part (iv) also shows that we have

$$T = \Sigma \, \lambda_n P_n$$

provided $|I - P_n| \leq 1 \ (n = 1, 2, \ldots)$.

(2)   It is not difficult to give examples for which P_n fails to be Hermitian; see Remark (1) following Theorem 29.5.

Our next goal is to show that equation (1) holds, with appropriate summability interpretation, for any compact Hermitian operator on a weakly complete space. We need first to recall some facts about almost periodic functions; see for example [114].

We denote by AP the Banach space of all complex valued uniformly almost periodic functions on $\underset{\sim}{R}$; AP is a subspace of $BC(\underset{\sim}{R})$, the bounded continuous functions on $\underset{\sim}{R}$. Given a function ϕ on $\underset{\sim}{R}$ and $a \in \underset{\sim}{R}$, we denote by ϕ_a the translate of ϕ defined by

$$\phi_a(t) = \phi(t + a) \qquad (t \in \underset{\sim}{R}).$$

Given $\phi \in C(\underset{\sim}{R})$ we have

$$\phi \in AP \iff \{\phi_a : a \in \underset{\sim}{R}\} \text{ is relatively compact in } BC(\underset{\sim}{R}). \tag{2}$$

Given ϕ differentiable on $\underset{\sim}{R}$ we have

$$\phi \text{ bounded}, \quad \phi' \in AP \implies \phi \in AP. \tag{3}$$

Each $\phi \in AP$ admits a Fourier transform defined by

$$c(\lambda) = \lim_{R \to \infty} \frac{1}{2R} \int_{-R}^{R} \phi(t) \, e^{-i\lambda t} dt \qquad (\lambda \in \underset{\sim}{R}), \tag{4}$$

and moreover $c(\lambda) = 0$ except on a countable set. The associated series

$$\sum_{\lambda \in \underset{\sim}{R}} c(\lambda) \, e^{i\lambda t}$$

is called the (generalized) Fourier series of ϕ. The Fourier transform $\{c(\lambda)\}$ satisfies the uniqueness condition

$$c(\lambda) = 0 \quad (\lambda \in \underset{\sim}{R}) \Rightarrow \phi = 0. \tag{5}$$

Finally, given $\phi \in AP$, let $\{\lambda_k\} = \{\lambda : c(\lambda) \neq 0\}$ and let (θ_{nk}) be the matrix of coefficients of the corresponding Fejér-Bochner kernel. Then

$$\lim_{n\to\infty} \sup_{t\in \underset{\sim}{R}} |\phi(t) - \sum_{k=1}^{k_n} \theta_{nk} c(\lambda_k) e^{i\lambda_k t}| = 0, \tag{6}$$

i. e. the Fourier series of ϕ is Fejér-Bochner summable to ϕ uniformly on $\underset{\sim}{R}$.

Lemma 2. Let $T \in B(X)$ be compact and Hermitian, let $x \in X$, $f \in X'$, and let

$$\phi(t) = f(\exp(itT)x) \qquad (t \in \underset{\sim}{R}).$$

Then $\phi \in AP$.

Proof. The function ϕ is bounded, since

$$|\phi(t)| \leq \|f\| \; |\exp(itT)| \; \|x\| \leq \|f\| \; \|x\| \qquad (t \in \underset{\sim}{R}),$$

and it is differentiable with

$$\phi'(t) = if(T \exp(itT)x) \qquad (t \in \underset{\sim}{R}).$$

We show now that $\{\phi'_a : a \in \underset{\sim}{R}\}$ is relatively compact in $BC(\underset{\sim}{R})$. Since

$$|f(\exp(itT)x)| \leq \|f\| \; \|x\| \qquad (t \in \underset{\sim}{R}),$$

it follows that the mapping $x \rightarrow \phi$ is a bounded linear mapping of X into $BC(\underset{\sim}{R})$. Note that

$$\phi'_a(t) = if(\exp(itT)T \exp(iaT)x).$$

Since T is compact and Hermitian, $\{T \exp(iaT)x : a \in \underset{\sim}{R}\}$ is relatively compact in X and hence $\{\phi'_a : a \in \underset{\sim}{R}\}$ is relatively compact in $BC(\underset{\sim}{R})$. We now have $\phi' \in AP$ by (2), and $\phi \in AP$ by (3).

Remark. The statement in the lemma clearly holds under the weaker hypothesis that $\{\exp(itT) : t \in \underset{\sim}{R}\}$ is bounded.

We show that, when X is weakly complete, in particular when X is reflexive, we can use the existence of the Fourier transform of ϕ to define the family of spectral projections for the operator T. In fact, for each $\lambda \in \underset{\sim}{R}$, by Lemma 2 and (4), we may define a linear operator E_λ on X by

$$f(E_\lambda x) = \lim_{R\to\infty} \frac{1}{2R} \int_{-R}^{R} f(\exp(itT)x)\, e^{-i\lambda t} dt \quad (x \in X,\ f \in X'), \qquad (7)$$

and then $|E_\lambda| \leq 1$.

Lemma 3. <u>Let</u> X <u>be weakly complete, let</u> $T \in B(X)$ <u>be compact and Hermitian and let</u> E_λ <u>be defined by</u> (7). <u>Then</u>

(i) $TE_\lambda = \lambda E_\lambda$,

(ii) $Tx = \lambda x \iff E_\lambda x = x$.

Proof. (i) Let $x \in X$, $f \in X'$. Then

$$\begin{aligned} f(TE_\lambda x) &= \lim_{R\to\infty} \frac{1}{2R} \int_{-R}^{R} f(T\exp(itT)x)\, e^{-i\lambda t}\, dt \\ &= \lim_{R\to\infty} \frac{1}{2iR} \int_{-R}^{R} e^{-i\lambda t} \frac{d}{dt} f(\exp(itT)x) dt \\ &= \lim_{R\to\infty} \frac{\lambda}{2R} \int_{-R}^{R} f(\exp(itT)x)\, e^{-i\lambda t}\, dt \\ &= f(\lambda E_\lambda x), \end{aligned}$$

which proves (i).

(ii) Let $x \in X$ be such that $Tx = \lambda x$. Then

$$\exp(itT)x = e^{i\lambda t} x,$$

and so $f(E_\lambda x) = f(x)$ for each $f \in X'$; $E_\lambda x = x$.

Now let $x \in X$ be such that $E_\lambda x = x$. By (i),

$$Tx = TE_\lambda x = \lambda E_\lambda x = \lambda x.$$

Corollary 4. <u>For each</u> $\lambda \in \underset{\sim}{R}$, <u>either</u> $E_\lambda = 0$, <u>or</u> λ <u>is an eigen-value of</u> T <u>and</u> E_λ <u>is the corresponding spectral projection at</u> λ.

Proof. For each $x \in X$, (i) gives $TE_\lambda x = \lambda E_\lambda x$, and then (ii) gives $E_\lambda(E_\lambda x) = E_\lambda x$. Therefore $E_\lambda^2 = E_\lambda$, and the result follows from (ii) and Theorem 1 (i).

Theorem 5. Let X be weakly complete, let $T \in B(X)$ be compact and Hermitian, and let λ_n, P_n be as in Theorem 1. Then X is the closed linear span of the eigenspaces of T, and T is the Fejér-Bochner sum of the series $\Sigma \lambda_n P_n$, i.e.

$$\lim_{n\to\infty} \left| T - \sum_{k=1}^{k_n} \theta_{nk}\lambda_k P_k \right| = 0.$$

Proof. Let $x \in X$, let $f \in X'$ be such that f annihilates each eigenspace of T (including $\ker(T)$), and let

$$\phi(t) = f(\exp(itT)x) \qquad (t \in \underset{\sim}{R}).$$

Corollary 4 gives $f(E_\lambda x) = 0$ $(\lambda \in \underset{\sim}{R})$, and so ϕ has zero Fourier transform by (7). Then $\phi = 0$ by (5), and so

$$f(x) = \phi(0) = 0.$$

Since x was arbitrary, this gives $f = 0$ and completes the proof that X is the closed linear span of the eigenspaces of T.

Let $Q_n = \sum_{k=1}^{k_n} \theta_{nk} E_{\lambda_k}$. For $x \in X$, $f \in X'$ we have

$$\begin{aligned} f(Q_n x) &= \sum_{k=1}^{k_n} \theta_{nk} f(E_{\lambda_k} x) \\ &= \sum_{k=1}^{k_n} \theta_{nk} c(\lambda_k) \\ &\to f(x) \text{ as } n \to \infty \end{aligned}$$

by (6). Therefore $Q_n \to I$ in the weak operator topology and $\{Q_n\}$ is bounded. For $x \in E_{\lambda_k} X$, we have $Q_n x = \theta_{nk} x$ for n sufficiently large. Since $\lim_{n\to\infty} \theta_{nk} = 1$, it follows that $\lim_{n\to\infty} \|Q_n x - x\| = 0$ for $x \in E_\lambda X$ and so for x in the linear span of the eigenspaces of T. Since this span is dense and $\{Q_n\}$ is bounded, we deduce that $Q_n \to I$ in the strong operator topology. But T is compact and $TQ_n = Q_nT$, so that $\lim_{n\to\infty} |TQ_n - T| = 0$. Lemma 3 (i) completes the proof.

Corollary 6. Under the hypotheses of Theorem 5, the least closed subalgebra of $B(X)$ that contains T coincides with the closed linear span of $\{P_n\}$.

Proof. The closed linear span of $\{P_n\}$ is an algebra and it contains T by the theorem. It is well known (see for example [119]) that each P_n belongs to the least closed subalgebra of $B(X)$ that contains T.

The conclusion of Corollary 6 holds equally under the hypotheses of Theorem 1 (iii) or (iv). Moreover, in the case of Theorem 1 (iii) the Vidav-Palmer theorem shows that the closed subalgebra of $B(X)$ generated by I and T is isometrically isomorphic to $C(Sp(T))$.

The example below shows that the statement of Theorem 5 does not hold for arbitrary complex Banach spaces X. Recall that c is the Banach space of all convergent complex sequences with the supremum norm.

Example 7. Let T be defined on c by

$$(Tx)(n) = \frac{1}{n}x(n) \qquad (n = 1, 2, 3, \ldots).$$

Then T is compact and Hermitian on c, but c is not the closed linear span of the eigenspaces of T.

Proof. It is easily verified that T is compact and Hermitian and that the closed linear span of the eigenspaces of T is c_0. In this case we still have $T = \Sigma \lambda_n P_n$.

29. EXAMPLES OF HERMITIAN ELEMENTS

We collect together in this section various examples of Hermitian elements. Most of the examples are given as operators, and we recall that $T \in B(X)$ has real spatial numerical range if and only if it has real numerical range as an element of any subalgebra of $B(X)$ that contains T and I.

The first series of examples centres around operators on Banach function spaces. We begin with function spaces on finite sets, i. e. $\underset{\sim}{C}^n$ with an absolute norm. The first two results are due to Tam [213].

Lemma 1. <u>Let $\|\cdot\|$ be an absolute norm on $\underset{\sim}{C}^n$, let $\lambda_1, \ldots, \lambda_n \in \underset{\sim}{R}$ and let T be defined on $\underset{\sim}{C}^n$ by</u>

$$(Tx)(r) = \lambda_r x(r) \qquad (1 \le r \le n).$$

<u>Then T is Hermitian.</u>

Proof. Given $t \in \underset{\sim}{R}$, we have

$$(\exp(itT)x)(r) = e^{it\lambda_r} x(r)$$

and so $\|\exp(itT)x\| = \|x\|$, since $\|\cdot\|$ is absolute. Therefore T is Hermitian.

The next result shows that for the case $n = 2$ the above real diagonal matrices are the only Hermitian operators unless the norm is an inner product norm, in which case the Hermitian operators correspond to the self-adjoint operators.

Lemma 2. <u>Let $\|\cdot\|$ be an absolute normalized norm on $\underset{\sim}{C}^2$, let</u>

$$T = \begin{bmatrix} a & b \\ c & d \end{bmatrix}$$

<u>be Hermitian and let $bc \neq 0$. Then $a, d \in \underset{\sim}{R}$, $b = c^*$ and $\|\cdot\|$ is the l_2-norm.</u>

Proof. We use the notation of Lemma 21. 5 and note from that result that $a, d \in \underset{\sim}{R}$. Since $\alpha z + \beta z^* \in \underset{\sim}{R}$ for all $|z| = 1$ if and only if $\beta = \alpha^*$, it follows that

$$(1 - t)(1 + (1 - t)\gamma/\psi(t))c = t(1 - t\gamma/\psi(t))b^* \qquad \cdot(1)$$

for $0 < t < 1$, $\gamma \in G(t)$. Suppose $\gamma_1, \gamma_2 \in G(t)$, $\gamma_1 > \gamma_2$, and let

$$p_j = (1 - t)\gamma_j/\psi(t), \qquad q_j = t\gamma_j/\psi(t) \qquad (j = 1, 2).$$

It follows from (1) that

$$\frac{1+p_1}{1-q_1} = \frac{1+p_2}{1-q_2},$$

and hence

$$(p_1 - p_2)(1 - q_2) + (q_1 - q_2)(1 + p_2) = 0.$$

Since $p_1 > p_2$, $q_1 > q_2$, and $|q_2| \leq 1$, $|p_2| \leq 1$ by Lemma 21.1, we deduce that $q_2 = 1$, $p_2 = -1$, and so $\gamma_2 = 0$. Similarly we have

$$(p_1 - p_2)(1 - q_1) + (q_1 - q_2)(1 + p_1) = 0,$$

which gives $\gamma_1 = 0$. This contradiction shows that ψ is differentiable for $0 < t < 1$. Using (1) again we obtain a constant k such that

$$\{t^2 + k(1-t)^2\}\psi'(t) = \{t - k(1-t)\}\psi(t) \qquad (0 < t < 1),$$

and therefore

$$\psi(t) = \theta\{t^2 + k(1-t)^2\}^{\frac{1}{2}} \qquad (0 < t < 1).$$

Since $\psi(0) = \psi(1) = 1$, we conclude that $\theta = k = 1$, and the norm is the l_2-norm.

It is clear, on taking l_1-direct sums of Hilbert spaces, that for $n \geq 3$ there exist non-diagonal Hermitian operators on $\underset{\sim}{C}^n$ when $\underset{\sim}{C}^n$ has absolute normalized norms other than the l_2-norm. Tam [213] shows that, if the norm on $\underset{\sim}{C}^n$ is also symmetric, i. e. invariant under permutations of $\{1, 2, \ldots, n\}$, then there exist non-diagonal Hermitian operators on $\underset{\sim}{C}^n$ if and only if the norm is the l_2-norm. More generally Tam [213] shows that this last result holds for sequence spaces. In particular an operator T on l_p $(1 \leq p \leq \infty, p \neq 2)$ is Hermitian if and only if it is of the form

$$(Tx)(n) = h(n)\,x(n) \qquad (n = 1, 2, \ldots),$$

where $h \in l_\infty$. This last result is also easily obtained by direct considerations of $\Pi(l_p)$.

Suppose, for example, that T is a Hermitian element of $B(l_\infty)$; let e_j denote the element of l_∞ with 1 in the j-th coordinate and 0 elsewhere, and let f_j denote the j-th coordinate functional, i. e.

$$f_j(x) = \xi_j \qquad (x = \{\xi_k\} \in l_\infty).$$

If $y \in l_\infty$, $\|y\| \leq 1$, and $f_j(y) = 0$, then $\|e_j + y\| = 1$, and $f_j \in D(e_j + y)$. Therefore

$$f_j(T(e_j + y)) \in V(T) \subset \underset{\sim}{R}.$$

Since this holds also if we replace y by $-y$ and by iy, it follows that $f_j(Te_j) = \lambda_j \in \underset{\sim}{R}$, and $f_j(Ty) = 0$. Moreover, by homogeneity, we have $f_j(Ty) = 0$ whenever $f_j(y) = 0$. Since $f_j(x - \xi_j e_j) = 0$ for all $x = \{x_k\} \in l_\infty$, we have $f_j(T(x - \xi_j e_j)) = 0$, i. e.

$$f_j(Tx) = \lambda_j \xi_j \qquad (x = \{\xi_k\} \in l_\infty).$$

Since this holds for all j, this gives

$$Tx = \{\lambda_k \xi_k\} \qquad (x = \{\xi_k\} \in l_\infty).$$

Conversely if T is of this form with $\{\lambda_k\}$ a bounded real sequence, then the method used in the proof of Lemma 1 shows that T is Hermitian.

Let E be a compact Hausdorff space and let $C(E)$ $(C_{\underset{\sim}{R}}(E))$ denote as usual the Banach space of continuous complex (real) functions on E. The following characterization of the Hermitian operators on $C(E)$ was given by G. Lumer in a lecture to the North British Functional Analysis Seminar in 1968.

Theorem 3. Let $T \in B(C(E))$. Then T is Hermitian if and only if it is of the form

$$Tf = hf \qquad (f \in C(E)),$$

where $h \in C_{\underset{\sim}{R}}(E)$.

Proof. Let T be Hermitian. Given $g \in C(E)$ with $g(x_0) = 1 = \|g\|$, the point evaluation at x_0 belongs to $D(g)$ and therefore $(Tg)(x_0) \in V(T) \subset \mathbb{R}$. In particular $h = T1 \in C_{\mathbb{R}}(E)$. Let $f \in C_{\mathbb{R}}(E)$, $\|f\| \leq 1$, $f(x_0) = 0$. Let $f = f^+ - f^-$ be the usual decomposition of f into its positive and negative parts. Then $f^+(x_0) = 0$ and $0 \leq f^+ \leq 1$. Therefore $(1 - f^+)(x_0) = 1 = \|1 - f^+\|$, which gives $(T(1 - f^+)(x_0) \in \mathbb{R}$. Therefore $(Tf^+)(x_0) \in \mathbb{R}$, and similarly $(Tf^-)(x_0) \in \mathbb{R}$, so that $(Tf)(x_0) \in \mathbb{R}$. Let

$$v(x) = \{1 - (f(x))^2\}^{\frac{1}{2}}, \qquad g(x) = v(x) + if(x) \qquad (x \in E).$$

Then $v(x_0) = g(x_0) = \|v\| = \|g\| = 1$, and hence

$$(Tv)(x_0) \in \mathbb{R}, \qquad (Tv)(x_0) + i(Tf)(x_0) \in \mathbb{R},$$

and therefore $(Tf)(x_0) = 0$. Finally, given $f \in C_{\mathbb{R}}(E)$, $x \in E$, we have $(f - f(x)1)(x) = 0$ and so

$$(Tf)(x) = f(x)(T1)(x) = h(x)\, f(x) \qquad (x \in E).$$

Therefore $Tf = hf$ $(f \in C(E))$, and the converse is immediate by the proof employed in Lemma 1.

Remarks. (1) Let E be locally compact and let $C_0(E)$ denote the Banach space of continuous complex functions on E that vanish at infinity. By modifying the above argument we can show that T is Hermitian on $C_0(E)$ if and only if it is a multiplication by some $h \in C_{\mathbb{R}}(E)$ (see Torrance [214]).

(2) By developing the above argument, Lumer [41] and Tam [213] characterize the Hermitian operators on various Orlicz spaces. In particular, if $X = L^p(S, \mu)$ where μ is non-atomic and totally σ-finite, then T is Hermitian on X if and only if it is of the form

$$Tf = hf \qquad (f \in X)$$

for some real valued $h \in L^{\infty}(S, \mu)$.

(3) Given that the Hermitian operators on a Banach function space X are multiplications, one can readily characterize the linear

isometries of X onto itself. We illustrate the technique with G. Lumer's proof of the Banach-Stone theorem; see also Lumer [41], Tam [213].

Theorem 4. *Let* U *be a linear isometry of* C(E) *onto itself. Then there is an automorphism* ϕ *of* C(E) *such that*

$$Uf = (U1)\,\phi(f) \qquad (f \in C(E)).$$

Proof. Given $h \in C_{\mathbb{R}}(E)$, let $T_h f = hf$ $(f \in C(E))$. Then T_h, UT_hU^{-1} are Hermitian and so by Theorem 3 there is $\phi(h) \in C_{\mathbb{R}}(E)$ such that $UT_hU^{-1} = T_{\phi(h)}$. It is easily verified that ϕ is an automorphism of $C_{\mathbb{R}}(E)$. Moreover with $f = U1$ we obtain

$$Uh = UT_h 1 = UT_hU^{-1}U1 = T_{\phi(h)}U1 = (U1)\phi(h) \qquad (h \in C_{\mathbb{R}}(E)).$$

Since U is linear on C(E), the proof is complete.

For the next family of Hermitian operators we return to the theme of almost periodic functions introduced in the previous section. Let K be a non-empty compact subset of $\mathbb{R}$, let E_0 be the linear span of $\{u_\lambda : \lambda \in K\}$ where

$$u_\lambda(s) = e^{i\lambda s} \qquad (s \in \mathbb{R}),$$

and let AP(K) be the closure of E_0 in the Banach space of bounded continuous functions on $\mathbb{R}$, i.e. AP(K) is the Banach space of all uniformly almost periodic functions with their Fourier exponents in K.

Theorem 5. *Each* $f \in AP(K)$ *is differentiable on* $\mathbb{R}$, *and if* T *is defined by* $Tf = -if'$ $(f \in AP(K))$, *then* T *is Hermitian.*

Proof. Let $T_0 f = -if'$ $(f \in E_0)$. Obviously $T_0 u_\lambda = \lambda u_\lambda$, and so

$$\exp(itT_0)u_\lambda(s) = \exp(it\lambda)u_\lambda(s) = u_\lambda(s+t) \qquad (s, t \in \mathbb{R}).$$

Let $\lambda_1, \ldots, \lambda_n \in K$, let E_1 be the linear span of $\{u_{\lambda_1}, \ldots, u_{\lambda_n}\}$, and let T_1 be the restriction of T_0 to E_1. We have

$$\exp(itT_1)\,f(s) = f(s+t) \qquad (f \in E_1,\ s,t \in \underset{\sim}{R})$$

and so $|\exp(itT_1)| = 1$, T_1 is Hermitian. Evidently $sp(T_1) = \{\lambda_1, \ldots, \lambda_n\}$, and so by Theorem 26.2

$$|T_1| = \rho(T_1) = \max\{|\lambda_1|, \ldots, |\lambda_n|\} \leq \tau = \max\{|\lambda| : \lambda \in K\}.$$

Since $\lambda_1, \ldots, \lambda_n$ are arbitrary, we have thus proved that $T_0 \in B(E_0)$ and $|T_0| = \tau$. Therefore T_0 has a unique extension to $T \in B(AP(K))$ and $|T| = \tau$.

Finally, given $f \in AP(K)$, there exist $f_n \in E_0$ uniformly convergent to f. Then $\{T_0 f_n\}$ converges uniformly to Tf, and so

$$\begin{aligned}
\int_0^t (Tf)(s)ds &= \lim_{n\to\infty} \int_0^t (T_0 f_n)(s)ds \\
&= \lim_{n\to\infty} -i \int_0^t f_n'(s)ds \\
&= \lim_{n\to\infty} -i(f_n(t) - f_n(0)) \\
&= -i(f(t) - f(0)).
\end{aligned}$$

This proves that f is differentiable on $\underset{\sim}{R}$ and that $Tf = -if'$. Since

$$\exp(itT)f(s) = f(s+t) \qquad (f \in AP(K),\ s,t \in \underset{\sim}{R}),$$

it is clear that T is Hermitian.

Remarks. (1) With $K = \{-1, 0, 1\}$, T corresponds to the Hermitian element h in NRI Example 6.1. Let P be the spectral projection corresponding to 1. Then P is not Hermitian and $|I - P| > 1$. To see this note that

$$P(\alpha u_{-1} + \beta u_0 + \gamma u_1) = \gamma u_1.$$

Consider the point $x = 6u_{-1} + 6u_0 - u_1$. We have

$$\|(I - P)x\| = \sup_\theta |6e^{-i\theta} + 6| = 12,$$

but

$$\|x\| = \sup_\theta |6e^{-i\theta} + 6 - e^{i\theta}|$$
$$= \sup_\theta |6 + 5\cos\theta - 7i\sin\theta|$$
$$= 11.$$

This shows that $|I - P| > 1$, and therefore, by Theorem 26.2, $I - P$ is not Hermitian, and so P is not Hermitian.

(2) When $K = [-1, 1]$, T is a slightly disguised form of the differentiation operator on $D([-1, 1])$, namely Browder's generator for the extremal algebra $Ea([-1, 1])$; see [125].

Let A be a unital C*-algebra. Recall that a linear operator D on A is a star derivation if

$$D(xy) = (Dx)y + xDy \qquad (x, y \in A)$$
$$Dx^* = -(Dx)^* \qquad (x \in A).$$

Sinclair [63] shows that $T \in B(A)$ is Hermitian if and only if it is of the form

$$Tx = hx + Dx \qquad (x \in A)$$

where $h^* = h$, and D is a star derivation; the proof uses several technical results about C*-algebras.

Finally we return to the Hermitian generator h of the extremal algebras $Ea(K)$ where $K \subset \mathbb{R}$. When $K = [-1, 1]$, Theorem 24.9 gives

$$V(h^2) = \{\phi''(0) : \phi \text{ entire}, \ \phi(0)=1, \ |\phi(z)| \le e^{|\mathrm{Re}\, z|} \quad (z \in \mathbb{C})\}. \quad (2)$$

We do not have an explicit description of the boundary of the above compact convex set, but we give below some of the known estimates for $V(h^2)$; see Bollobás [117], and [134].

Theorem 6. For the algebra $Ea([-1, 1])$,

$$[-\tfrac{i}{8}, \tfrac{i}{8}] \cup \{z : |z - \tfrac{1}{2}| \le \tfrac{1}{2}\} \subset V(h^2) \subset \{z : \mathrm{Re}\, z \ge 0\}.$$

Proof. Given $f \in D(1)$, we have $\mathrm{Re}\, f(\cos th) \le 1$ $(t \in \underset{\sim}{R})$ and so

$$1 - \tfrac{1}{2}t^2\, \mathrm{Re}\, f(h^2) + \chi(t) \le 1 \qquad (t \in \underset{\sim}{R})$$

where $\chi(t) = 0(t^4)$ as $t \to 0$. Therefore $\mathrm{Re}\, f(h^2) \ge 0$.

Given $|\zeta| \le 1$, let

$$\phi(z) = \tfrac{1}{2}(1 - \zeta) + \tfrac{1}{2}(1 + \zeta) \cosh z \qquad (z \in \underset{\sim}{C}).$$

Then ϕ is entire, $\phi(0) = 1$, $|\phi(z)| \le 2e^{|z|}$ and

$$|\phi(it)| = \tfrac{1}{2}|(1 + \cos t) - \zeta(1 - \cos t)| \le 1 \qquad (t \in \underset{\sim}{R}).$$

We now have $|\phi(z)| \le e^{|\mathrm{Re}\, z|}$ $(z \in \underset{\sim}{C})$ by Phragmen-Lindelöf. Since $\phi''(0) = \tfrac{1}{2}(1 + \zeta)$, it follows from (2) that

$$\{z : |z - \tfrac{1}{2}| \le \tfrac{1}{2}\} \subset V(h^2).$$

Finally let

$$\psi(z) = 1 \pm \tfrac{1}{2}i(\cosh(\tfrac{1}{2}z) - 1) - \frac{1}{4}(\cosh(\tfrac{1}{2}z) - 1)^2 \qquad (z \in \underset{\sim}{C}).$$

Then ψ is entire, $\psi(0) = 1$, $|\psi(z)| \le 2e^{|z|}$, $\psi''(0) = \pm i/8$. To complete the proof, it is now sufficient to show that

$$|\psi(it)| \le 1 \qquad (t \in \underset{\sim}{R}).$$

This is equivalent to showing that for $0 \le \beta \le 1$, we have

$$|1 \pm i\beta - \beta^2| \le 1$$

i.e.

$$(1 - \beta^2)^2 \le 1 - \beta^2,$$

and this last inequality is obvious.

Remark. B. Bollobás (private communication) has shown that $\pm\frac{1}{4}i \in V(h^2)$.

For the Hermitian generator h of $Ea([0,1])$ it is easy to show by the above arguments that $V(h^2)$ has interior, but in this case $V(h^2)$ contains no open interval of the imaginary axis.

Theorem 7. For the algebra $Ea([0,1])$, $V(h^2) \cap i\mathbb{R} = \{0\}$.

Proof. Let $k = 2h - 1$, and then $V(k) = [-1, 1]$. Let $f \in D(1)$ and suppose $f(h^2) = ic$ with $c \in \mathbb{R}$. Then

$$\frac{1}{4}ic = 1 + 2f(k) + f(k^2).$$

We note that $\gamma = -1 - 2f(k) \in \mathbb{R}$ and $f(k^2) = \gamma + \frac{1}{4}ic$. Since $|f(\exp(itk))| \leq 1$ $(t \in \mathbb{R})$ we have

$$|1 + itf(k) - (\gamma + \tfrac{1}{4}ic)\tfrac{1}{2}t^2 + 0(t^3)| \leq 1. \tag{3}$$

By considering (3) for small t we deduce that

$$f(k)^2 \leq \gamma = -1 - 2f(k) ,$$

and therefore $f(k) = -1$. Theorem 26.9 now gives $f(k^2) = 1$, $c = 0$.

30. THE RUSSO-DYE-PALMER THEOREM

In the proof of the Vidav-Palmer theorem in Volume I of these notes, NRI Theorem 6.9, we quoted without proof a technical lemma about B*-algebras. To be precise, let A be a unital B*-algebra and let $E = \{\exp(ih) : h \in A, h^* = h\}$. Then the closed unit ball of A is the closed convex hull of E. This result is Palmer's extension of a result of Russo and Dye [58]. The main aim of this section is to give an elementary proof of Palmer's result; the proof is due to Harris [160] and is based upon the use of generalized Möbius transformations (first introduced by Potapov [192]). This elementary proof of NRI Lemma 6.8 thus completes a proof of the Vidav-Palmer theorem in which only elementary Banach algebra techniques are employed. (A simple proof of the commutative case has been given by Burckel [128].)

Throughout this section A denotes a complex unital B*-algebra, Δ denotes the closed unit disc in $\mathbb{C}$, and Γ the unit circle in $\mathbb{C}$. We write

$$U = \{u \in A : uu^* = u^*u = 1\}$$
$$E = \{\exp(ih) : h \in A, h^* = h\}.$$

Let $x \in A$, $\|x\| < 1$. Then $\rho(x^*) = \rho(x) < 1$, and $\rho(xx^*)=\rho(x^*x)= \|x\|^2 < 1$. We may therefore define F by

$$F(\lambda) = (1-xx^*)^{-\frac{1}{2}}(\lambda+x)(1+\lambda x^*)^{-1}(1-x^*x)^{\frac{1}{2}} \quad (|\lambda| < \|x\|^{-1}). \quad (1)$$

Lemma 1. <u>Given</u> $x \in A$, $\|x\| < 1$, <u>the mapping</u> F <u>defined by</u> (1) <u>is analytic and</u>

$$F(\lambda) \in U \qquad (\lambda \in \Gamma).$$

Proof. It is routine that F is analytic on the given neighbourhood of Δ. Let $\lambda \in \Gamma$. We note the following identities:

$$(1 + \lambda x^*)^{-1}(\lambda + x) = x + \lambda(1 + \lambda x^*)^{-1}(1 - x^*x) \quad (2)$$

$$(\lambda + x)(1 + \lambda x^*)^{-1} = x + \lambda(1 - xx^*)(1 + \lambda x^*)^{-1} \quad (3)$$

$$(1 - xx^*)^{\frac{1}{2}}x = x(1 - x^*x)^{\frac{1}{2}}. \quad (4)$$

Equations (2), (3) are elementary and (4) follows from a consideration of the power series expansion of $(1 - y)^{\frac{1}{2}}$ when $\|y\| < 1$. Since $\rho(x) < 1$, $F(\lambda)$ is invertible and, by (2), (3), (4),

$$\begin{aligned}(F(\lambda)^{-1})^* &= (1 - xx^*)^{\frac{1}{2}}(\lambda^* + x^*)^{-1}(1 + \lambda^*x)(1 - x^*x)^{-\frac{1}{2}} \\ &= (1 - xx^*)^{\frac{1}{2}}(1 + \lambda x^*)^{-1}(\lambda+x)(1 - x^*x)^{-\frac{1}{2}} \\ &= (1 - xx^*)^{-\frac{1}{2}}\{x + \lambda(1 - xx^*)(1 + \lambda x^*)^{-1}\}(1 - x^*x)^{\frac{1}{2}} \\ &= F(\lambda).\end{aligned}$$

This shows that $F(\lambda)$ is unitary for each $\lambda \in \Gamma$.

Theorem 2. (Russo-Dye.) $A_1 = \overline{\text{co}}\, U$.

Proof. Let $x \in A$, $\|x\| < 1$ and let F be given by (1). Since F is analytic we have

$$F(0) = \frac{1}{2\pi}\int_0^{2\pi} F(e^{i\theta})\, d\theta.$$

By (4), $F(0) = x$, and then by Lemma 1, $x \in \overline{\text{co}}\, U$. The result follows easily.

Theorem 3. (Palmer.) $A_1 = \overline{\text{co}}\, E$.

Proof. By Theorem 2, it is enough to show that $U \subset \overline{\text{co}}\, E$. Given $u \in U$, $0 < t < 1$, let $x = tu$. Then x is normal, and it is easily verified that

$$(\lambda + F(\lambda))(1 + \lambda x^*) = 2\lambda(1 + \tfrac{1}{2}(\lambda^* x + \lambda x^*)) \quad (\lambda \in \Gamma).$$

Since $\|x\| < 1$, it follows that $\lambda + F(\lambda))$ is invertible and so $-\lambda \notin \mathrm{Sp}(F(\lambda))$. Therefore $\mathrm{Sp}(F(\lambda))$ is a proper compact subset of Γ and so we may define $h = -i \log F(\lambda)$, and we have $h^* = h$, since $F(\lambda) \in U$. Thus $F(\lambda) = \exp(ih) \in E$ for $\lambda \in \Gamma$, and it follows as in Theorem 2 that $x \in \overline{\text{co}}\, E$. Therefore $U \subset \overline{\text{co}}\, E$, and the proof is complete.

The conclusion of Theorem 3 can be sharpened.

Theorem 4. (Palmer [189].) $\{x \in A : \|x\| < 1\} \subset \text{co}\, E$.

Proof. Let $h \in A$, $h^* = h$, $\|h\| < 1$. Then $h = \tfrac{1}{2}(u + v)$, where

$$u = h + i(1 - h^2)^{\frac{1}{2}}, \qquad v = h - i(1 - h^2)^{\frac{1}{2}}.$$

It is easily verified that $u, v \in U$, $1 \notin \mathrm{Sp}(u)$, $1 \notin \mathrm{Sp}(v)$, and hence $u, v \in E$, $h \in \text{co}\, E$. Let $x \in A$, $\|x\| < \tfrac{1}{2}$. Then $x = h + ik$, where $h^* = h$, $k^* = k$, $\|h\| < \tfrac{1}{2}$, $\|k\| < \tfrac{1}{2}$. It follows that $x \in \text{co}\, E$. Finally let $x \in A$, $\|x\| < 1$, and let $t = \|x\|^{-1}$. Then $tx \in \overline{\text{co}}\, E$ by Theorem 3, and so there exists $y \in \text{co}\, E$ such that

$$\|tx - y\| < \tfrac{1}{2}(t - 1).$$

By the above, $tx - y = (t - 1)z$ for some $z \in \text{co}\, E$. Therefore

$$x = t^{-1}y + (1 - t^{-1})z \in \text{co}\, E.$$

Remarks. (1) Harris [160] employs the generalized Möbius transformations to obtain related results on more general Banach star algebras; see also Pták [193], [194].

(2) Let T be a compact Hausdorff space and let $f \in C(T)$, $\|f\| < 1$. Then f is a convex combination of functions in $C(T)$ which

take their values in the unit circle. This is the continuous analogue of the theorem of Fisher [152] for the disc algebra, which is used in §32.

31. THE LINEAR SPANS OF THE STATES

In this section we discuss the complex and real linear spans of the set of states on a unital normed algebra. The results, which constitute an interesting contribution to the geometry of normed algebras, are due to Moore [186] and Sinclair [203]. They include in particular a characterization (due to R. T. Moore) of B*-algebras in terms of the structure of the dual space of the algebra.

Let A denote a complex unital normed algebra, and $D(1) = D(A, 1)$ the set of normalized states on A, i.e.

$$D(1) = \{f \in A' : \|f\| = f(1) = 1\}.$$

We recall (NRI Theorem 4.1) that the numerical radius v, which is given by

$$v(a) = \sup\{|f(a)| : f \in D(1)\},$$

is a norm on A equivalent to the given algebra-norm $\|\cdot\|$, in fact that

$$\frac{1}{e}\|a\| \leq v(a) \leq \|a\| \qquad (a \in A).$$

We denote also by v the norm on A' dual to v, i.e.

$$v(f) = \sup\{|f(a)| : v(a) \leq 1\} \qquad (f \in A').$$

Plainly v is a norm on A' equivalent to the usual norm, and with $v(f) \geq \|f\|$.

The Bohnenblust-Karlin theorem (NRI Theorem 4.5) states that $D(1)$ is a total subset of A', and so the complex linear span of $D(1)$ is weak* dense in A'. The following theorem shows that in fact the complex linear span of $D(1)$ is the whole of A'.

Theorem 1. A' *is the complex linear span of* $D(1)$. *Moreover, if* $f \in A'$, *then there exist* $\alpha_k \in \mathbb{R}^+$, $f_k \in D(1)$ $(k = 1, 2, 3, 4)$ *with*

$$\alpha_1 + \alpha_2 + \alpha_3 + \alpha_4 \le \sqrt{2}\, v(f),$$

$$f = \alpha_1 f_1 - \alpha_2 f_2 + i(\alpha_3 f_3 - \alpha_4 f_4).$$

Proof. Denote by C the space of all continuous complex functions on the compact Hausdorff space $D(1)$ (with the weak* topology). The mapping T defined by

$$(Ta)(\phi) = \phi(a) \qquad (\phi \in D(1)),$$

is a linear isometry of A with the norm v into C with the uniform norm, and $T1$ is the constant function 1. Therefore T^* is a linear mapping of C' into A' with

$$v(T^*g) \le \|g\| \qquad (g \in C'),$$

and maps the set P of probability measures into $D(1)$.

Given $f \in A'$, the functional g_0, defined on TA by

$$g_0(Ta) = f(a),$$

can be extended to an element g of C' with $\|g\| = v(f)$ and $T^*g = f$. The measure g can be decomposed into the form

$$g = \alpha_1 g_1 - \alpha_2 g_2 + i(\alpha_3 g_3 - \alpha_4 g_4)$$

with $g_k \in P$, $\alpha_k \in \underset{\sim}{R}^+$ $(k = 1, 2, 3, 4)$, and with

$$\alpha_1 + \alpha_2 + \alpha_3 + \alpha_4 \le \sqrt{2}\, \|g\|.$$

Finally,

$$f = T^*g = \alpha_1 T^*g_1 - \alpha_2 T^*g_2 + i(\alpha_3 T^*g_3 - \alpha_4 T^*g_4),$$

is the required decomposition of f.

An interesting direct proof of Theorem 1 has been given by Sinclair [203], based on the following lemma.

Lemma 2. Let E be a finite subset of $\underset{\sim}{C}$, let Δ denote the closed unit disc $\{\zeta \in \underset{\sim}{C} : |\zeta| \le 1\}$, let $\eta \ge 0$, and suppose that

$\eta\Delta \subset \text{co } E$. <u>Then</u> co(ED(1)) <u>is a weak* compact convex subset of</u> A' <u>and</u>

$$f \in A', \quad v(f) \le \eta \implies f \in \text{co}(ED(1)) .$$

Proof. Here $ED(1) = \{\beta f : \beta \in E, f \in D(1)\}$. Let $E = \{\beta_1, \dots, \beta_n\}$, and let $F = \{\underset{\sim}{\alpha} = (\alpha_1, \dots, \alpha_n) \in \underset{\sim}{R}^{+n} : \alpha_1 + \alpha_2 + \dots + \alpha_n = 1\}$. Since each set $\beta_k D(1)$ is compact in the weak* topology, the set

$$\beta_1 D(1) \times \beta_2 D(1) \times \dots \times \beta_n D(1) \times F$$

is compact in the product topology, and co(ED(1)) is the image of this set under the mapping

$$(\beta_1 f_1, \beta_2 f_2, \dots, \beta_n f_n, \underset{\sim}{\alpha}) \to \alpha_1 \beta_1 f_1 + \dots + \alpha_n \beta_n f_n .$$

Thus co(ED(1)) is a weak* compact convex subset of A'. Let $f \in A'$ with $v(f) \le \eta$. If $f \notin \text{co}(ED(1))$, then the Hahn-Banach separation theorem, applied to the locally convex space A' with the weak* topology, gives the existence of $a \in A$ and $\varepsilon > 0$ such that

$$\text{Re } f(a) - \varepsilon \ge \sup \{\text{Re } g(a) : g \in \text{co}(ED(1))\}.$$

Given $g \in D(1)$, choose $\zeta \in \underset{\sim}{C}$ with $|\zeta| = 1$ such that $|g(a)| = \text{Re}(\zeta g(a))$. Then $\eta\zeta \in \text{co } E$, and so there exists $\underset{\sim}{\alpha} \in F$ with $\eta\zeta = \alpha_1\beta_1 + \dots + \alpha_n\beta_n$. Then $\eta\zeta g \in \text{co}(ED(1))$, and therefore

$$\eta\,|g(a)| = \text{Re}(\eta\zeta g(a)) \le \text{Re } f(a) - \varepsilon.$$

Since this holds for all $g \in D(1)$, we have $\text{Re } f(a) > \eta v(a)$, contradicting the assumption that $v(f) \le \eta$.

Second proof of Theorem 1. Take $E = \{1, -1, i, -i\}$. Then $\frac{1}{\sqrt{2}}\Delta \subset \text{co } E$, and so given $f \in A'$ with $v(f) \le \frac{1}{\sqrt{2}}$, there exist $\alpha_k \in \underset{\sim}{R}^+$, $f_k \in D(1)$ $(k = 1, 2, 3, 4)$ such that $\alpha_1 + \alpha_2 + \alpha_3 + \alpha_4 = 1$ and $f = \alpha_1 f_1 - \alpha_2 f_2 + i(\alpha_3 f_3 - \alpha_4 f_4)$.

Definition 3. We denote by $H(A')$ the real linear span of $D(1)$, i.e.

$$H(A') = \{\alpha f - \beta g : f, g \in D(1), \ \alpha, \beta \in \underset{\sim}{R}^+\}.$$

The elements of $H(A')$ are called Hermitian functionals.

Corollary 4. $A' = H(A') + iH(A')$; and, given $f \in A'$, there exist $g, h \in H(A')$ with $f = g + ih$ and

$$\|g\| + \|h\| \le \sqrt{2}\, v(f).$$

Proof. In the notation of Theorem 1, we take $g = \alpha_1 f_1 - \alpha_2 f_2$, $h = \alpha_3 f_3 - \alpha_4 f_4$. Plainly $\|g\| + \|h\| \le \alpha_1 + \alpha_2 + \alpha_3 + \alpha_4 \le \sqrt{2}\, v(f)$.

Definition 5. Given $h \in H(A')$, we define $\|h\|_H$ by

$$\|h\|_H = \inf\{\alpha+\beta : \alpha f - \beta g = h, \ \alpha, \beta \in \underset{\sim}{R}^+, \ f, g \in D(1)\}.$$

It is clear from the definition that

$$\|h\| \le \|h\|_H \qquad (h \in H(A')),$$

and so $\|\cdot\|_H$ is a norm on the real linear space $H(A')$.

We base the proof of the next theorem on two elementary lemmas on upper semi-continuous set valued mappings.

Lemma 6. *Let X, Y, Z be topological spaces of which Y is compact, let f be a continuous mapping of $X \times Y$ into Z, and let $F(x) = f(x, Y) = \{f(x, y) : y \in Y\}$ $(x \in X)$. Then F is upper semi-continuous.*

Proof. Let $x_0 \in X$ and let U be a neighbourhood of $F(x_0)$. Given $y_0 \in Y$, there exist open neighbourhoods $V(y_0)$, $W(y_0)$ of x_0, y_0 respectively such that

$$(x, y) \in V(y_0) \times W(y_0) \Rightarrow f(x, y) \in U.$$

Choose such $V(y_0)$, $W(y_0)$ for each $y_0 \in Y$. Then, by compactness of

Y, there exist $y_1, \ldots, y_n$ with $Y \subset \bigcup_{k=1}^{n} W(y_k)$. Let $V = \bigcap_{k=1}^{n} V(y_k)$. Then V is a neighbourhood of x_0, and

$$x \in V,\ y \in Y \Rightarrow (x, y) \in V(y_k) \times W(y_k) \text{ for some } k,$$
$$\Rightarrow f(x, y) \in U.$$

Thus $F(x) \subset U \quad (x \in V)$.

Lemma 7. Let F be an upper semi-continuous mapping of a topological space X into the subsets of a Hausdorff space Z. For each $z_0 \in Z$, the set $\{x \in X : z_0 \in F(x)\}$ is closed.

Proof. Straightforward.

Theorem 8. (i) $\|\cdot\|_H$ is a norm on H(A') and $\|h\| \le \|h\|_H$ $(h \in H(A'))$.

(ii) Given $h \in H(A')$, there exist $\alpha, \beta \in \underset{\sim}{R}^+$, $f, g \in D(1)$ with $h = \alpha f - \beta g$ and $\|h\|_H = \alpha + \beta$.

(iii) H(A') with the norm $\|\cdot\|_H$ is a real Banach space.

Proof. We have seen that (i) is an immediate consequence of the definition of $\|\cdot\|_H$. To prove (ii), let $h \in H(A')$ with $\|h\|_H < 1$. An application of Lemma 6 with $X = [0,1] \times [0,1]$ and $Y = D(1) \times D(1)$ shows that the set valued mapping

$$(\alpha, \beta) \to \alpha D(1) - \beta D(1) = \{\alpha f - \beta g : (f, g) \in D(1) \times D(1)\}$$

is upper semi-continuous from $[0,1] \times [0,1]$ to the subsets of A' with the weak* topology. Then Lemma 7 shows that the set

$$E = \{(\alpha, \beta) \in [0,1] \times [0,1] : h \in \alpha D(1) - \beta D(1)\}$$

is closed and therefore compact. It is also non-void, since $\|h\|_H < 1$. Therefore the function $(\alpha, \beta) \to \alpha + \beta$ attains its minimum on E, and (ii) is proved.

To prove (iii), let $h_n \in H(A')$ with

$$\|h_{n+1} - h_n\|_H < 2^{-n} \qquad (n = 1, 2, \ldots).$$

By (i), it follows that $h_1 + \sum_{n=1}^{\infty} (h_{n+1} - h_n)$ converges in the Banach space $(A', \|\cdot\|)$ to an element h. By (ii), there exist $\alpha_n, \beta_n \in \underset{\sim}{R}^+$, $f_n, g_n \in D(1)$ such that

$$h_{n+1} - h_n = \alpha_n f_n - \beta_n g_n, \qquad \|h_{n+1} - h_n\|_H = \alpha_n + \beta_n.$$

With convergence with respect to $\|\cdot\|$, we have

$$h - h_m = \sum_{n=m}^{\infty} (h_{n+1} - h_n) = \sum_{n=m}^{\infty} (\alpha_n f_n - \beta_n g_n). \tag{1}$$

Let

$$\gamma_m = \sum_{n=m}^{\infty} \alpha_n, \qquad \delta_m = \sum_{n=m}^{\infty} \beta_n \qquad (m = 1, 2, \ldots).$$

If $\gamma_m \neq 0$, let $\phi_m = \gamma_m^{-1} \sum_{n=m}^{\infty} \alpha_n f_n$, and if $\gamma_m = 0$, let ϕ_m be an arbitrary element of D(1). Similarly let $\psi_m = \delta_m^{-1} \sum_{n=m}^{\infty} \beta_n g_n$ if $\delta_m \neq 0$, and otherwise let ψ_m be an arbitrary element of D(1). Since D(1) is convex and closed with respect to $\|\cdot\|$, $\phi_m, \psi_m \in D(1)$, and (1) gives

$$h - h_m = \gamma_m \phi_m - \delta_m \psi_m. \tag{2}$$

Thus $h - h_m \in H(A')$, $h \in H(A')$. Also (2) gives

$$\|h - h_m\|_H \leq \gamma_m + \delta_m$$
$$= \sum_{n=m}^{\infty} (\alpha_n + \beta_n) = \sum_{n=m}^{\infty} \|h_{n+1} - h_n\|_H < \frac{1}{2^{m-1}}$$

Thus $\lim_{m \to \infty} \|h - h_m\|_H = 0$, and the proof is complete.

Remark. The numbers α, β in (ii) are uniquely determined by h, for we have

$$\alpha + \beta = \|h\|_H, \quad \alpha - \beta = \alpha f(1) - \beta g(1) = h(1).$$

NRI Theorem 6.9 (Vidav-Palmer) characterizes B*-algebras among unital Banach algebras as those for which there are enough elements of

H(A). R. T. Moore has pointed out that they may also be characterized as those for which there are not too many elements of $H(A')$. The proof of this depends on the following lemma.

Lemma 9. Let B be a closed subalgebra of a normed star algebra C, and given $f \in C'$, let f^* be defined by $f^*(x) = (f(x^*))^*$ $(x \in C)$. Then the following statements are equivalent.

(i) B is a star subalgebra of C.

(ii) $f \in C'$, $f(B) = \{0\} \Rightarrow f^*(B) = \{0\}$.

Proof. That (i) $\Rightarrow$ (ii) is obvious. Suppose then that $b \in B$ but $b^* \notin B$. By the Hahn-Banach theorem, there exists $f \in C'$ with $f(B) = \{0\}$ but $f(b^*) \neq 0$. Therefore $f^*(B) \neq \{0\}$, (ii) fails.

Theorem 10. Let A be complete. Then the following statements are equivalent.

(i) $H(A') \cap iH(A') = \{0\}$.

(ii) $H(A) + iH(A) = A$.

(iii) A admits an involution with respect to which it is a B^*-algebra.

Proof. That (ii) is equivalent to (iii) is NRI Theorem 6.9. If $*$ is an involution on A with respect to which A is a B^*-algebra, then $a \in H(A)$ if and only if $a^* = a$, and consequently (by Dixmier [140] 2.6.4 (p. 40))

$$\begin{aligned} H(A') &= \{f \in A' : f(H(A)) \subset \mathbb{R}\} \\ &= \{f \in A' : f(a^*) = (f(a))^* \quad (a \in A)\}, \end{aligned} \tag{3}$$

from which (i) follows at once.

It only remains to prove that (i) implies (ii). Suppose that $H(A') \cap iH(A') = \{0\}$. Then, by Corollary 4,

$$A' = H(A') \oplus iH(A'). \tag{4}$$

Thus each $f \in A'$ has a unique expression in the form

$$f = \mathrm{re}\, f + i.\, \mathrm{im}\, f,$$

with re f, im f $\in$ H(A'). (re f should not be confused with Re f defined by (Re f)(a) = Re(f(a)) (a $\in$ A).) By Corollary 4,

$$\|\mathrm{re}\, f\| \leq \sqrt{2}\, v(f) \leq \sqrt{2}\, \|f\|,$$

and so f → re f is a bounded real linear projection of A' onto H(A') (with respect to the usual norm $\|\cdot\|$). We define an involution f → f* on A' by taking

$$f^* = \mathrm{re}\, f - i.\, \mathrm{im}\, f \quad (f \in A');$$

and we have proved that f → f* is continuous on (A', $\|\cdot\|$).

Consider the Banach algebra A" with the Arens multiplication. By NRI Theorem 12.2, for F $\in$ A" we have

$$F \in H(A'') \iff \{F(f) : f \in D(1)\} \subset \underset{\sim}{R}. \tag{5}$$

Given F $\in$ A", we define F* by F*(f) = (F(f*))* (f $\in$ A'). By continuity of the mapping f → f*, we have F* $\in$ A", and evidently

$$\tfrac{1}{2}(F + F^*)(f),\ \frac{1}{2i}(F - F^*)(f) \in \underset{\sim}{R} \qquad (f \in H(A')).$$

Since D(1) $\subset$ H(A'), (5) now gives

$$\tfrac{1}{2}(F + F^*),\ \frac{1}{2i}(F - F^*) \in H(A''),$$

and so

$$A'' = H(A'') + iH(A'').$$

Therefore, by NRI Theorem 6.9, A" with the involution F → F* just constructed is a B*-algebra.

Let a → $\hat{a}$ denote the canonical mapping of A into A". Then $\hat{A}$ is a closed subalgebra of A". It is enough to prove that $\hat{A}$ is a star subalgebra of A". For then $\hat{A}$ is a B*-algebra, and since the canonical mapping is an isometric isomorphism, we have A = H(A) + iH(A).

If $\hat{A}$ is not a star subalgebra of A", then, by Lemma 9, there exists $\Phi \in A'''$ with $\Phi(\hat{A}) = \{0\}$ but $\Phi^*(\hat{A}) \neq \{0\}$. Since A" is a B*-algebra, (3) shows that $\Phi = \Phi_1 + i\Phi_2$ with

$$\Phi_k(F^*) = (\Phi_k(F))^* \qquad (F \in A'', \ k = 1, 2). \tag{6}$$

Then we have $\Phi_1 = \alpha_1\Psi_1 - \alpha_2\Psi_2$ with $\alpha_1, \alpha_2 \in \underset{\sim}{R}^+$ and $\Psi_1, \Psi_2 \in D(A'', \hat{1})$. Clearly $\Psi_k|_{\hat{A}} \in D(\hat{A}, \hat{1})$ $(k = 1, 2)$, and so $\Phi_1|_{\hat{A}} \in H(\hat{A}')$. Similarly $\Phi_2|_{\hat{A}} \in H(\hat{A}')$. By the isometric isomorphism between A and $\hat{A}$, we have

$$H(\hat{A}') \cap iH(\hat{A}') = \{0\}.$$

Since $\Phi_1|_{\hat{A}} + i\Phi_2|_{\hat{A}} = \Phi|_{\hat{A}} = 0$, it follows that $\Phi_1|_{\hat{A}} = \Phi_2|_{\hat{A}} = 0$. But then, by (6),

$$\Phi_k(\hat{A}^*) = \{0\} \qquad (k = 1, 2),$$

and so $\Phi(\hat{A}^*) = \{0\}$, a contradiction.

Remark. Moore has given a further proof of Theorem 10 by using Theorem 12 below.

Theorem 11. The closed unit ball in $H(A')$ with respect to the norm $\|\cdot\|_H$ is compact in the weak* topology.

Proof. Let $U = \{h \in H(A') : \|h\|_H \le 1\}$, and let

$$K = \{(f, g, \alpha, \beta) \in D(1) \times D(1) \times [0,1] \times [0,1] : \alpha + \beta \le 1\}.$$

With the weak* topology on each of the sets $D(1)$, K is compact in the product topology. Also, by Theorem 8, U is the image of K under the mapping T given by

$$T(f, g, \alpha, \beta) = \alpha f - \beta g.$$

Therefore U is compact in the weak* topology.

Theorem 12. Let A be complete. Then the following statements are equivalent.

(i) $H(A')$ is $\|\cdot\|$ closed in A'.

(ii) $\|\cdot\|_H$ is equivalent to $\|\cdot\|$ on $H(A')$.

(iii) $H(A')$ is weak* closed in A'.

(iv) $H(A') = \{f \in A' : f(H(A)) \subset \underset{\sim}{R}\}$.

Proof. We have $\| \cdot \| \leq \| \cdot \|_H$ and $H(A')$ is complete with respect to $\| \cdot \|_H$ (Theorem 8). Therefore, by Banach's isomorphism theorem (i) $\Rightarrow$ (ii), and obviously (ii) $\Rightarrow$ (i). It is also obvious that (iv) $\Rightarrow$ (iii) $\Rightarrow$ (i). Suppose that $H(A')$ is weak* closed in A' and let $R(A') = \{f \in A' : f(H(A)) \subset \underset{\sim}{R}\}$. Evidently $H(A') \subset R(A')$. Suppose that $f \in R(A') \backslash H(A')$. Then there exists $a \in A$ and $\varepsilon > 0$ such that

$$\text{Re } f(a) - \varepsilon \geq \sup \text{Re} \{g(a) : g \in H(A')\}.$$

Since $H(A')$ is a real linear space, this gives $\text{Re } g(a) = 0$ $(g \in H(A'))$ and so $\text{Im } g(ia) = 0$ $(g \in D(1), ia \in H(A))$. But then $f(ia) \in \underset{\sim}{R}$, and so $\text{Re } f(a) = 0$ which is contradictory. Thus $H(A') = R(A')$ and (iii) $\Rightarrow$ (iv).

It only remains to prove that (ii) $\Rightarrow$ (iii). With $r > 0$, let

$$A'_r = \{f \in A' : \|f\| \leq r\}, \qquad U_r = \{h \in H(A') : \|h\|_H \leq r\},$$

and assume that $\| \cdot \|_H$ is equivalent to $\| \cdot \|$ on $H(A')$. Then, given $r > 0$, there exists $s > 0$ such that $A'_r \cap H(A') \subset U_s$, and so

$$A'_r \cap U_s \subset A'_r \cap H(A') \subset A'_r \cap U_s.$$

Therefore $A'_r \cap H(A') = A'_r \cap U_s$, which is weak* compact. Since this holds for every $r > 0$, the Krein-Šmulian theorem (Dunford and Schwartz [148] Theorem V. 5. 7, p. 429) shows that $H(A')$ is weak* closed in A'. Thus (ii) $\Rightarrow$ (iii), and the proof is complete.

Example 13. Let A be the disc algebra of all complex valued functions continuous on the unit disc Δ that are analytic in its interior, and let $R(A') = \{f \in A' : f(H(A)) \subset \underset{\sim}{R}\}$. For this Banach algebra A we have $R(A') \neq H(A')$, and so $H(A')$ is not norm closed in A'. To see this, note first that the evaluation functionals δ_ζ given by $\delta_\zeta a = a(\zeta)$ $(a \in A)$ belong to $D(1)$ for each $\zeta \in \Delta$. Thus all elements of $H(A)$ are real valued on Δ, and are therefore real multiples of the unit element.. Therefore

$$R(A') = \{f \in A' : f(1) \in \underset{\sim}{R}\},$$

from which

$R(A') \cap iR(A') = \{f \in A' : f(1) = 0\}$.

Next regard A as a subalgebra of the algebra $C(\Gamma)$ of all continuous complex functions on the unit disc Γ. Let $f \in H(A') \cap iH(A')$, and let μ, ν be real measures on Γ such that $f = \mu|_A = i\nu|_A$. Then

$$\int_\Gamma e^{in\theta}\, d\mu(\theta) = i \int_\Gamma e^{in\theta}\, d\nu(\theta) \qquad (n = 0, 1, 2, \ldots);$$

and complex conjugation gives

$$\int_\Gamma e^{-in\theta}\, d\mu(\theta) = -i \int_\Gamma e^{-in\theta}\, d\nu(\theta) \qquad (n = 0, 1, 2, \ldots).$$

Therefore, by the F. and M. Riesz theorem, $\mu + i\nu$ is absolutely continuous with respect to Lebesgue measure. Since the real part of an absolutely continuous measure is absolutely continuous, μ is absolutely continuous.

We have proved that if $f \in H(A') \cap iH(A')$, then each extension of f to a real measure on Γ is absolutely continuous with respect to Lebesgue measure. Finally, take a non-zero real singular measure μ on Γ with $\mu(1) = 0$. Then $\mu|_A \in R(A') \cap iR(A')$, but $\mu|_A \notin H(A') \cap iH(A')$.

We are grateful to R. T. Moore for permission to include the following theorem which he gave in his lecture at the Conference on Numerical Ranges (Aberdeen, July 1971). The theorem shows in particular that if A is complete but does not admit a B* involution, then $D(1)$ contains a non-trivial copy of the unit disc $\Delta = \{\zeta \in \mathbb{C} : |\zeta| \leq 1\}$.

Theorem 14. <u>Suppose that</u> $H(A') \cap iH(A') \neq \{0\}$. <u>Then there exist</u> $f \in D(1)$ <u>and</u> $g \in \frac{1}{4}(D(1) - D(1))$ <u>with</u> $g \neq 0$, <u>such that</u> $f + \Delta g \subset D(1)$.

Proof. Since $H(A') \cap iH(A') \neq \{0\}$, there exist $\alpha_k \in \mathbb{R}^+$, $f_k \in D(1)$ $(k = 1, 2, 3, 4)$ with

$$0 \neq \alpha_1 f_1 - \alpha_2 f_2 = i(\alpha_3 f_3 - \alpha_4 f_4).$$

Evaluation of the functionals at 1, gives $\alpha_1 - \alpha_2 = i(\alpha_3 - \alpha_4)$, and so $\alpha_1 = \alpha_2$, $\alpha_3 = \alpha_4$. By renumbering if necessary we may suppose that we have β with $0 < \beta \leq 1$ such that

$$i(f_3 - f_4) = \beta(f_1 - f_2) \neq 0.$$

Let $f = \frac{1}{4}(f_1 + f_2 + f_3 + f_4)$, $g = \frac{1}{4}(f_3 - f_4)$. With $\zeta = \xi + i\eta \in \Delta$, $\xi, \eta \in \underset{\sim}{R}$, we have

$$\begin{aligned} f + \zeta g &= \tfrac{1}{4}(f_1 + f_2 + f_3 + f_4) + \tfrac{1}{4}\xi(f_3 - f_4) + \tfrac{1}{4}\beta\eta(f_1 - f_2) \\ &= \tfrac{1}{4}(1 + \beta\eta)f_1 + \tfrac{1}{4}(1 - \beta\eta)f_2 + \tfrac{1}{4}(1 + \xi)f_3 + \tfrac{1}{4}(1 - \xi)f_4. \end{aligned}$$

Since $\xi, \beta\eta \in [-1, 1]$, the coefficients of f_1, f_2, f_3, f_4 are all non-negative, and they add up to 1 since $(f + \zeta g)(1) = f(1) = 1$. Thus $f + \zeta g \in D(1)$ $(\zeta \in \Delta)$.

Remark. Another result in this general area, given in Browder [125] Theorem 8. 6, is that any point derivation on a function algebra is a scalar multiple of the difference between two normalized states.

32. NUMERICAL INDEX

We recall that the <u>numerical index</u> of a normed algebra A is defined by

$$n(A) = \inf\{v(a) : a \in A, \|a\| = 1\}.$$

Given a normed space X, we denote n(B(X)) more simply by n(X), and we call n(X) the <u>numerical index</u> of the normed space X. For simplicity of exposition we shall suppose throughout this section that all spaces and algebras are over the complex field. Thus $e^{-1} \leq n(A) \leq 1$. We show first that the range of the numerical index is the whole of the closed interval $[e^{-1}, 1]$.

We recall that Ψ is the family of continuous convex functions ψ on $[0, 1]$ such that

$$\max(1-t,\ t) \leq \psi(t) \leq 1 \qquad (0 \leq t \leq 1)$$

and by Proposition 21. 3, Ψ is in 1-1 correspondence with the family of absolute normalized linear norms on $\underset{\sim}{C}^2$.

Lemma 1. Let $\psi \in \Psi$ with corresponding norm $\|\cdot\|$ on $\underset{\sim}{C}^2$ and let S be the shift operator on $\underset{\sim}{C}^2$ defined by

$$S(\alpha, \beta) = (0, \alpha).$$

Then $\|S\| = 1$, $V(S) = \{z : |z| \le v(S)\}$, and

$$v(S) = \sup_{0<t<1} (1 - t)\{1 + (1 - t)\frac{\psi_R'(t)}{\psi(t)}\}.$$

Proof. By Lemma 21.2, we have $|\alpha| \le \|(\alpha,\beta)\|$. Since also $S(1, 0) = (0, 1)$, it follows that $\|S\| = 1$. Proposition 21.5 gives

$$V(S) = \{(1 - t)[1 + (1 - t)\frac{\gamma}{\psi(t)}]\, z^* : (t, \gamma) \in \Xi,\ |z| = 1\}.$$

The formula for $v(S)$ follows by an application of NRI Theorem 9.3.

Let $J = [1, e]$. Given $r \in J$ let ψ_r be defined on $[0,1]$ by

$$\psi_r(t) = \begin{cases} (1 - t)\exp(\frac{\log r}{r}\,\frac{t}{1-t}) & 0 \le t \le \frac{r}{1+r} \\ t & \frac{r}{1+r} < t \le 1. \end{cases}$$

It is easily verified that ψ_r is continuous and that $\psi_r''(t) \ge 0$ $(t \in (0,1)\backslash r(r + 1)^{-1})$, so that ψ_r is convex. It is now clear that $\psi_r \in \Psi$. Let $\|\cdot\|_r$ be the corresponding norm on $\underset{\sim}{C}^2$. Given a linear operator T on $\underset{\sim}{C}^2$, we write $\|T\|_r$, $v_r(T)$ respectively for the operator norm and numerical radius determined by $\|\cdot\|_r$. We write $n_r(\underset{\sim}{C}^2)$ for the numerical index of $\underset{\sim}{C}^2$ with respect to $\|\cdot\|_r$.

Lemma 2. $n_1(\underset{\sim}{C}^2) = 1$, $n_e(\underset{\sim}{C}^2) = e^{-1}$.

Proof. Since $\|\cdot\|_1$ is the l_∞-norm on $\underset{\sim}{C}^2$ the first assertion follows from Theorem 5 below. In fact, in this simple case, if

$$T = \begin{bmatrix} a & b \\ c & d \end{bmatrix},$$

it is elementary to verify (with the help of Proposition 21.5) that

$$v_1(T) = \max\{|a| + |b|,\ |c| + |d|\} = \|T\|_1 .$$

With S as in Lemma 1, we have by direct calculation that

$$v(S) = \max[\tfrac{1}{c}, \sup\{\tfrac{1-t}{t} : \tfrac{e}{c+1} \le t < 1\}] = \tfrac{1}{e}.$$

Therefore $n_e(\underset{\sim}{C}^2) = e^{-1}$ by Lemma 1.

Lemma 3. Given compact metric spaces A, B and a continuous real function F on $A \times B$, let G, H be defined on A by

$$G(x) = \sup\{F(x,y) : y \in B\}, \qquad H(x) = \inf\{F(x,y) : y \in B\}.$$

Then G, H are continuous on A.

Proof. We note that F is uniformly continuous since $A \times B$ is compact, and an elementary argument then shows that G, H are continuous.

Theorem 4. $\{n_r(\underset{\sim}{C}^2) : r \in J\} = [e^{-1}, 1]$.

Proof. By Lemma 2, it is sufficient to show that the mapping $r \to n_r(\underset{\sim}{C}^2)$ is continuous on J. We denote the matrix of $T \in B(\underset{\sim}{C}^2)$ by

$$T = \begin{bmatrix} a & b \\ c & d \end{bmatrix}.$$

Let $K = \{T \in B(\underset{\sim}{C}^2) : |a| + |b| + |c| + |d| = 1\}$. Since $\underset{\sim}{R}^+ K = B(\underset{\sim}{C}^2)$ we have

$$n_r(\underset{\sim}{C}^2) = \inf\{v_r(T)/\|T\|_r : T \in K\}.$$

Since K is compact, it is now sufficient by Lemma 3 to show that the real functions

$$(r, T) \to v_r(T), \qquad (r, T) \to \|T\|_r$$

are continuous on $J \times K$.

Proposition 21.5 and direct computation show that $v_r(T)$ is the maximum of $r^{-1}|c| + |d|$ and

$$\sup\{|(a+btzr)(1-t\log r)+(cz^*+rdt)r^{-1}\log r| : 0 \le t \le 1,\ |z| = 1\}.$$

An application of Lemma 3 now gives the continuity of $(r, T) \to v_r(T)$.

Let $L = \{(\alpha, \beta) \in \underset{\sim}{C}^2 : |\alpha| + |\beta| = 1\}$. Then

$$\|T\|_r = \sup\{\|Tx\|_r / \|x\|_r : x \in L\}.$$

To show that $(r, T) \to \|T\|_r$ is continuous it is now sufficient by Lemma 3 to show that the mappings

$$(r, T, x) \to \|Tx\|_r, \qquad (r, T, x) \to \|x\|_r$$

are continuous on $J \times K \times L$. This last step is elementary.

We give next some classes of Banach spaces with numerical index 1. The proof given in Theorem 5 below is due to C. M. McGregor.

Theorem 5. For any compact Hausdorff space E, we have $n(C(E)) = 1$.

Proof. Let $T \in B(C(E))$, $\|T\| = 1$, and let $\varepsilon > 0$. Then we can choose $f \in C(E)$ with $\|f\| = 1$ and $u \in E$ such that

$$f(u) \neq 0, \qquad |(Tf)(u)| > 1 - \varepsilon.$$

Now choose an open neighbourhood U of u with $0 \notin f(U)$. By Urysohn's lemma there is a continuous function ϕ from E to $[0,1]$ with $\phi(u) = 1$, $\phi(E\backslash U) = \{0\}$. Now define p on E by $p(t) = 0$ when $f(t) = 0$ and

$$p(t) = f(t)\, |f(t)|^{-1} \{1 - |f(t)|^2\}^{\frac{1}{2}}\, \phi(t)$$

when $f(t) \neq 0$. Then $p \in C(E)$. Let $g = f + ip$, $h = f - ip$, so that $g, h \in C(E)$, $f = \frac{1}{2}g + \frac{1}{2}h$, and

$$|g(u)| = |h(u)| = \|g\| = \|h\| = 1.$$

Taking $F(k) = k(u)$ $(k \in C(E))$, we have $g(u)^*F \in D(C(E), g)$, $h(u)^*F \in D(C(E), h)$. Since either $|(Tg)(u)| > 1 - \varepsilon$, or $|(Th)(u)| > 1 - \varepsilon$, it follows that $v(T) > 1 - \varepsilon$. The result follows.

Lemma 6. If A, B are isometrically isomorphic normed algebras, then $n(A) = n(B)$; in particular, if X, Y are isometrically isomorphic normed spaces, then $n(X) = n(Y)$.

Proof. Elementary.

Remarks. (1) It is not true that the numerical index is monotonic with respect to norm-decreasing isomorphisms. To see this, let $A = Ea(K)$ with K the closed unit disc, and let $B = Ea(K)$ with the spectral norm. Then the identity map is a norm-decreasing isomorphism from A to B. We have $n(A) = e^{-1}$ by Corollary 24.11 and $n(B) = 1$ by Theorem 5. Now let S be the shift operator on $\mathbb{C}^2$. Let A be the algebra $\{\alpha I + \beta S : \alpha, \beta \in \mathbb{C}\}$ with the l_1 operator norm and let B be the same algebra with the l_2 operator norm $\|\cdot\|_2$. It is elementary to verify that $\|S\|_2 = 1$ and so

$$\|\alpha I + \beta S\|_2 \leq |\alpha| + |\beta| = \|\alpha I + \beta S\|_1.$$

Thus the identity map is again a norm-decreasing isomorphism from A to B, but in this case we have $n(A) = 1$, $n(B) = \frac{1}{2}$.

(2) The situation is similar for quotient maps. Take $A = M^{\omega}(\mathbb{C})$ where $\omega(\lambda) = e^{|\lambda|}$. Then A is an L-space and so $n(A) = 1$ (see Theorem 8 below). Let

$$I = \{\mu \in A : \int e^{z\lambda}\, d\mu(\lambda) = 0 \quad (|z| \leq 1)\}.$$

Then I is a closed ideal in A and $A/I = B$ (say) is the extremal algebra $Ea(K)$, so that $n(B) = e^{-1}$. Now let J be a maximal ideal in B. Then $B/J \approx \mathbb{C}$ and so $n(B/J) = 1$. Browder [125] shows that if A is a function algebra and I a closed ideal in A, then $n(A/I) \geq \frac{1}{2}$, and gives an example with $n(A/I) = \frac{1}{2}$. It follows in particular that the semi-simple commutative Banach algebra $Ea(K)$ is not the quotient of a function algebra.

Lemma 7. For any normed space X we have $n(X') \leq n(X)$. Thus $n(X') = n(X)$ if X is reflexive.

Proof. The map $T \to T^*$ is a linear isometry from $B(X)$ into $B(X')$ with respect to the operator norm and also with respect to the numerical radius (NRI Corollary 9. 6). The rest is clear.

Recall that a (complex) M-space with order unit is isometrically isomorphic to $C(E)$ for some compact Hausdorff space E. Also, the dual of an L-space is an M-space with order unit.

Theorem 8. Let X be an M-space or an L-space. Then $n(X) = 1$.

Proof. Let X be an M-space with order unit. Then $n(X) = 1$ by Theorem 5 and Lemma 6. Let X be an L-space. Then X' is an M-space with order unit, and, by Lemma 7, $1 \geq n(X) \geq n(X') = 1$. Finally let X be any M-space. Then X' is an L-space, so that $n(X') = 1$, $n(X) = 1$.

The list of spaces with numerical index 1 thus includes the classical spaces c_0, c, l_1, l_∞, $C(E)$, $M(E)$ and $L_1(S, \Sigma, \mu)$. We can now add the disc algebra to this list; the result is due to M. J. Crabb (see [135]).

Theorem 9. The disc algebra (as a Banach space) has numerical index 1.

Proof. Let X denote the disc algebra, let $T \in B(X)$ with $\|T\| = 1$ and let $\varepsilon > 0$. Choose $f \in X$ with $\|f\| = 1$ such that

$$\|Tf\| > 1 - \varepsilon.$$

By a theorem of Fisher [152], there exist $\alpha_1, \ldots, \alpha_n$ with $\alpha_j \geq 0$, $\alpha_1 + \ldots + \alpha_n = 1$, and finite Blaschke products $g_1, \ldots, g_n$ such that

$$\|f - \sum_{j=1}^{n} \alpha_j g_j\| < \varepsilon.$$

For some j with $1 \leq j \leq n$, we have $\|Tg_j\| > 1 - 2\varepsilon$. Choose $|z| = 1$ such that $|Tg_j(z)| = \|Tg_j\|$, and define ϕ on X by

$$\phi(f) = g_j(z)^* f(z) \qquad (f \in X).$$

Then $\phi \in D(X, g_j)$, and so $(Tg_j)(z)\, g_j(z)^* \in V(T)$. This gives $v(T) > 1 - 2\varepsilon$, and the result follows.

For finite dimensional spaces X, McGregor [181] shows that $n(X) = 1$ if and only if $|f(x)| = 1$ whenever x is an extreme point of the unit ball of X and f is an extreme point of the unit ball of X'. In particular $\underset{\sim}{C}^4$ has index 1 with respect to the norm

$$\|(a, b, c, d)\| = \max\{|a| + |b| + |c|, |d|\}.$$

With the above norm $\underset{\sim}{C}^4$ is neither an L-space nor an M-space. One can also show that a norm on $\underset{\sim}{C}^2$ has index 1 if and only if it is the l_∞ norm with respect to some change of basis. We show below that a similar uniqueness result holds for index e^{-1}. The next two results are due to Crabb [131], [130].

Theorem 10. Let $T \in B(X)$, $\|T\| = ev(T)$, $T \neq 0$ and let $p(T) = 0$ for some non-zero polynomial p. Then 0 is an eigenvalue of T of ascent greater than 1.

Proof. Let $v(T) = 1$, so that $\|\exp(\lambda T)\| \leq e^{|\lambda|}$ $(\lambda \in \underset{\sim}{C})$. Let $x \in S(X)$, $f \in S(X')$. Then

$$\sum_{n=0}^{\infty} \frac{|f(T^n x)|^2}{n!^2} = \frac{1}{2\pi} \int_0^{2\pi} |f(\exp(e^{i\theta}T)x)|^2 d\theta \leq e^2.$$

Since $\|T\| = e$, given $\varepsilon > 0$ there exist x and f such that $|f(Tx)|^2 \geq e^2 - \varepsilon^2$. It follows that

$$|f(T^n x)| \leq n!\,\varepsilon \qquad (n \neq 1).$$

Suppose that $a_0 + a_1 T + \ldots + a_n T^n = 0$. Then

$$\begin{aligned} |a_1 f(Tx)| &= |a_0 f(x) + a_2 f(T^2 x) + \ldots + a_n f(T^n x)| \\ &\leq \varepsilon(|a_0| + 2!\,|a_2| + \ldots + n!\,|a_n|). \end{aligned}$$

Therefore $a_1 = 0$, and, since $a_0 T + \ldots + a_n T^{n+1} = 0$, we also have $a_0 = 0$. The result follows.

Remark. If the condition $\|T\| = e\,v(T)$ is replaced by the condition $\|T^k\| = k!\,(\frac{e}{k})^k\,v(T)^k$, the above technique gives 0 an eigenvalue of T of ascent greater than k.

It is easily checked that the norm $\|\cdot\|_e$ on $\underset{\sim}{C}^2$ (as defined before Lemma 2) is specified by

$$\|(1, z)\| = \begin{cases} e^{|z|} & \text{if } |z| \leq 1 \\ e|z| & \text{if } |z| \geq 1. \end{cases}$$

Recall also that the index of $(\underset{\sim}{C}^2, \|\cdot\|_e)$ is attained on the shift operator. We are now ready for the converse.

Proposition 11. Let dim $X = 2$ and let $T \in B(X)$ with $v(T) = 1$, $\|T\| = e$. Then $T^2 = 0$ and there exists $x \in S(X)$ such that

$$\|I + zT\| = \|x + zTx\| = \begin{cases} e^{|z|} & \text{if } |z| \leq 1 \\ e|z| & \text{if } |z| \geq 1. \end{cases}$$

Proof. We have $T^2 = 0$ by the Cayley-Hamilton theorem and Theorem 10. Since $v(T) = 1$ we have

$$\|I + zT\| = \|\exp(zT)\| \leq e^{|z|} \qquad (z \in \underset{\sim}{C}).$$

Choose $x \in S(X)$, $f \in S(X')$ such that $f(Tx) = e$. We have $f(x) = 0$, by the proof of the above theorem, and so

$$\|(I + zT)x\| \geq |f((I + zT)x)| = e|z| \qquad (z \in \underset{\sim}{C}).$$

It follows that

$$\|I + zT\| = \|x + zTx\| = e \qquad (|z| = 1).$$

Given $|z| < 1$, choose $w \in \underset{\sim}{C}$ such that

$$|z + w| = |z| + |w| = 1.$$

Then

$$e = \|(I + (z + w)T)x\| = \|(I + wT)(I + zT)x\| \leq e^{|w|}\|x + zTx\|,$$

and so

and so

$$\|x + zTx\| \geq e^{|z|} \qquad (|z| < 1).$$

It follows that

$$\|(I + zT)x\| = \|I + zT\| = e^{|z|} \qquad (|z| < 1).$$

Given $|z| > 1$, let $z = a + b$ with $|a| = 1$, $|b| = |z| - 1$. Then

$$\begin{aligned}\|I + zT\| &\leq \|I + aT\| + |b| \, \|T\| \\ &= e + (|z| - 1)e = e|z|.\end{aligned}$$

The proof is now complete.

The power inequality fails in the extremal algebra $Ea(K)$ of Corollary 24.10 and the numerical index of $Ea(K)$ is e^{-1}. The power inequality clearly holds for algebras of numerical index 1. On the other hand, there exists norms on $\underset{\sim}{C}^2$ with numerical index arbitrarily close to 1 for which the power inequality fails; see [135].

33. SPECTRAL OPERATORS

The concept of spectral operator as introduced by Dunford [146] may be regarded as an axiomatization of the bounded operators which admit a satisfactory spectral decomposition. Berkson [6] and Lumer [42] showed how spectral operators of scalar type are related to Hermitian operators and Lumer [42] used numerical range techniques to establish the necessity of one of Dunford's axioms for a satisfactory theory of spectral operators. We give here an exposition of these ideas. We also include the recent extension of Fuglede's theorem, due to Berkson, Dowson and Elliott, and its application to a long standing problem on scalar operators.

Notation. Let X be a complex Banach space and let Γ be a total linear subspace of X', i. e.

$$f(x) = 0 \quad (f \in \Gamma) \Rightarrow x = 0.$$

Let Σ denote the σ-field of Borel subsets of $\underset{\sim}{C}$, and let $P(X)$ denote the set of projections (idempotents) in $B(X)$.

Definition 1. A mapping $E : \Sigma \to P(X)$ is called a spectral measure of class (Σ, Γ) if

(i) $E(\delta_1 \cup \delta_2) = E(\delta_1) + E(\delta_2) - E(\delta_1)E(\delta_2) \quad (\delta_1, \delta_2 \in \Sigma)$,

(ii) $E(\delta_1 \cap \delta_2) = E(\delta_1)E(\delta_2) \quad (\delta_1, \delta_2 \in \Sigma)$,

(iii) $E(\underset{\sim}{C} \setminus \delta) = I - E(\delta) \quad (\delta \in \Sigma)$,

(iv) $E(\underset{\sim}{C}) = I$,

(v) there is $M > 0$ such that $|E(\delta)| \leq M \quad (\delta \in \Sigma)$,

(vi) the mapping $\delta \to f(E(\delta)x)$ is countably additive on Σ for each $x \in X$, $f \in \Gamma$.

Definition 2. An operator $T \in B(X)$ is a prespectral operator of class Γ if

(i) there is a spectral measure E of class (Σ, Γ) such that

$$T\,E(\delta) = E(\delta)T \qquad (\delta \in \Sigma),$$

(ii) $\quad Sp(T|_{E(\delta)X}) \subset \bar{\delta} \qquad (\delta \in \Sigma)$.

The spectral measure E is then called a resolution of the identity of class Γ for T. An operator $T \in B(X)$ is a spectral operator if it is prespectral of class X'.

Definition 3. A prespectral operator $S \in B(X)$ with a resolution of the identity E of class Γ is called a scalar-type operator of class Γ if

$$S = \int_{Sp(S)} \lambda \, E(d\lambda).$$

(The integral is a Riemann-Stieltjes integral in the uniform operator topology; see Dunford [146] for details.)

Theorem 4. (Berkson.) *Let $S \in B(X)$ be a scalar-type operator of class Γ, and let E be a corresponding resolution of the identity. Then $S = R + iJ$ with $RJ = JR$, where*

$$R = \int_{Sp(S)} \mathrm{Re}\,\lambda \, E(d\lambda), \qquad J = \int_{Sp(S)} \mathrm{Im}\,\lambda \, E(d\lambda),$$

<u>and there is an equivalent norm</u> $|\cdot|$ <u>on</u> X <u>with respect to which</u> $R^m J^n$ <u>is Hermitian for</u> $m, n = 0, 1, 2, \ldots$.

Proof. Given $x \in X$, $f \in X'$, the variation of the finitely additive measure $f(E(\cdot)x)$ is defined by

$$\operatorname{var} f(E(\cdot)x) = \sup \Sigma |f(E(\delta_j)x)|,$$

where the supremum is taken over all finite sequences $\{\delta_j\}$ of pairwise disjoint Borel subsets of $\underset{\sim}{C}$. Given such a finite sequence $\{\delta_j\}$ we have

$$\begin{aligned}\Sigma |\operatorname{Re} f(E(\delta_j)x)| &= \Sigma' \operatorname{Re} f(E(\delta_j)x) - \Sigma'' \operatorname{Re} f(E(\delta_j)x)\\ &= \operatorname{Re} f(\Sigma' E(\delta_j)x) - \operatorname{Re} f(\Sigma'' E(\delta_j)x)\\ &\leq 2M \|f\| \|x\|,\end{aligned}$$

and hence $\operatorname{var} f(E(\cdot)x) \leq 4M\|f\| \|x\|$. We now define $|\cdot|$ on X by

$$|x| = \sup\{\operatorname{var} f(E(\cdot)x) : f \in X', \|f\| = 1\}.$$

It is readily verified that $|\cdot|$ is a seminorm on X, and also $|x| \leq 4M\|x\|$ $(x \in X)$. Given $x \in X$, choose $f \in X'$ such that $\|f\| = 1$, $f(x) = \|x\|$. Then

$$\|x\| = f(x) = f(E(\underset{\sim}{C})x) \leq |x| ,$$

and hence $|\cdot|$ is a norm on X equivalent to $\|\cdot\|$.

We show next that each $E(\delta)$ is Hermitian with respect to the norm $|\cdot|$. Let $\alpha \in \underset{\sim}{R}$, $x \in X$, $f \in X'$, $\|f\| = 1$, and let $\{\delta_j\}$ be a finite sequence of pairwise disjoint Borel subsets of $\underset{\sim}{C}$. Then

$$\begin{aligned}&\sum_{j=1}^{n} |f(E(\delta_j)(I + i\alpha E(\delta))x)|\\ &= \sum_{j=1}^{n} |f(E(\delta_j)[E(\underset{\sim}{C}\backslash\delta)x + (1 + i\alpha)E(\delta)x])|\\ &\leq \sum_{j=1}^{n} |f(E(\delta_j \cap (\underset{\sim}{C}\backslash\delta))x)| + |1 + i\alpha| \sum_{j=1}^{n} |f(E(\delta_j \cap \delta)x)|\\ &\leq |1 + i\alpha| \operatorname{var} f(E(\cdot)x).\end{aligned}$$

It follows that $|(I + i\alpha E(\delta))x| \leq |1 + i\alpha|\ |x|$ $(x \in X)$. Given $y \in E(\delta)X$, $|y| = 1$, we have

$$|(I + i\alpha E(\delta))y| = |y + i\alpha y| = |1 + i\alpha|.$$

Therefore $|I + i\alpha E(\delta)| = |1 + i\alpha|$, and $E(\delta)$ is Hermitian (with respect to $|\cdot|$) by NRI Lemma 5.2.

Since

$$S = \int_{Sp(S)} \lambda\ E(d\lambda),$$

it is clear that $S = R + iJ$ with $RJ = JR$. For $m, n = 0, 1, 2, \ldots$, we have

$$R^m J^n = \int_{Sp(S)} (\mathrm{Re}\ \lambda)^m (\mathrm{Im}\ \lambda)^n E(d\lambda).$$

Since $H(B(X))$ is a closed real linear subspace, it follows that each $R^m J^n$ is Hermitian with respect to $|\cdot|$.

Remarks. (1) With the notation of Theorem 4, Berkson [6] observes that the closed subalgebra of $B(X)$ generated by I, R and J is bicontinuously isomorphic to $C(Sp(S))$; when X has the norm $|\cdot|$, this isomorphism is in fact an isometry by the Vidav-Palmer theorem.

(2) Let A be a closed subalgebra of $B(X)$ containing I. Dunford [146] shows that, if A is bicontinuously isomorphic to $C(\Omega)$, for some compact Hausdorff space Ω, then, for each $S \in A$, S^* is a scalar-type operator of class X.

(3) When X is reflexive, the following statements are equivalent for $S \in B(X)$.

(i) S is a scalar-type operator of class X'.

(ii) $S = R + iJ$, where $RJ = JR$, and for an equivalent norm $|\cdot|$ on X, $R^m J^n$ is Hermitian $(m, n = 0, 1, 2, \ldots)$.

(4) Panchapagesan [52] gives the related polar decomposition of scalar-type operators.

We examine next in more detail the situation in which an operator or a family of operators can be made Hermitian by an equivalent renorming.

Definition 5. A subset Λ of $B(X)$ is <u>Hermitian-equivalent</u> if there is an equivalent norm on X for which each operator in Λ is Hermitian.

Definition 6. A subset Λ of a complex unital Banach algebra A is <u>pre-Hermitian</u> if there is an equivalent norm on A for which each element of Λ is Hermitian.

The two notions coincide when A is a closed subalgebra of $B(X)$; this result is due to R. T. Moore following a lecture of E. Berkson at the Conference on Numerical Ranges held in Aberdeen, July 1971.

Lemma 7. <u>Let A be a closed subalgebra of $B(X)$ containing I. A subset Λ of A is a Hermitian-equivalent subset of $B(X)$ if and only if it is a pre-Hermitian subset of A.</u>

Proof. Let Λ be pre-Hermitian in A. Then there is an equivalent norm $\|\cdot\|_0$ on A such that $\|\exp(itT)\|_0 = 1$ $(T \in \Lambda,\ t \in \mathbb{R})$. Hence the subgroup G of $G(A)$ generated by $\{\exp(itT) : T \in \Lambda,\ t \in \mathbb{R}\}$ is in the closed unit ball of $(A, \|\cdot\|_0)$ and so is bounded in $(A, \|\cdot\|)$. Let $|x| = \sup\{\|Sx\| : S \in G\}$ $(x \in X)$. By NRI Lemma 10.3 $|\cdot|$ is a norm on X equivalent to $\|\cdot\|$ and $\exp(itT)$ has norm one in the corresponding operator norm whenever $T \in \Lambda$, $t \in \mathbb{R}$. This shows that Λ is Hermitian-equivalent in $B(X)$, and the converse is obvious.

It is natural to ask if a family of operators on X is Hermitian-equivalent if and only if each member of the family is Hermitian-equivalent. The solution below for commutative families is due to Lumer [42].

Lemma 8. <u>Let Λ be a commutative subset of $B(X)$. Then Λ is Hermitian-equivalent if and only if each operator in the closed real linear span of Λ is Hermitian-equivalent.</u>

Proof. Let Y be the closed real linear span of Λ and suppose each $T \in Y$ is Hermitian-equivalent. For $k = 1, 2, 3, \ldots$, let

$$Y_k = \{T \in Y : |\exp(itT)| \le k \quad (t \in \mathbb{R})\}.$$

We have $Y = \cup\{Y_k : k = 1, 2, 3, \ldots\}$ by hypothesis. We show that each Y_k is closed. Let $T_n \in Y_k$, $T_n \to T$, and let $t \in \mathbb{R}$. Then $\exp(itT_n) \to \exp(itT)$. Hence $|\exp(itT)| \le k$ and so $T \in Y$. By Baire's category theorem some Y_k contains a ball of Y. For any $S \in Y$ we may now choose $U, V \in Y_k$, $r \ge 0$ such that $S = U + rV$. Then, since Λ is commutative, we have

$$|\exp(itS)| \le |\exp(itU)|\,|\exp(itrV)| \le k^2 \qquad (t \in \mathbb{R}).$$

It follows as in the proof of Lemma 7 that Y, and so Λ, is Hermitian-equivalent. The converse is clear.

In order to establish that a scalar-type operator of class Γ has a unique resolution of the identity of class Γ, we need first an extension of Fuglede's theorem (see e.g. Halmos [30], page 99).

Lemma 9. (Berkson, Dowson and Elliott [113].) *Let A be a complex unital Banach algebra and let $x, u, v \in A$ where u, v are commuting pre-Hermitians. If x commutes with $u + iv$, then x commutes with u and v.*

Proof. By Lemmas 7 and 8 we may suppose without loss that u and v are Hermitian. Let $p = u + iv$, $q = u - iv$. As in the proof of NRI Theorem 3.10 it is enough to show that the mapping

$$\lambda \to \exp(-\lambda q)x\exp(\lambda q) \qquad (\lambda \in \mathbb{C})$$

is bounded. Since $px = xp$ and $pq = qp$ we have

$$\begin{aligned}\exp(-\lambda q)x\exp(\lambda q) &= \exp(-\lambda q)\exp(\lambda^* p)x\exp(-\lambda^* p)\exp(\lambda q)\\ &= \exp(-\lambda q + \lambda^* p)x\exp(-\lambda^* p + \lambda q)\\ &= \exp(-ih)x\exp(ih)\end{aligned}$$

where $h = 2(\mathrm{Im}\,\lambda)u - 2(\mathrm{Re}\,\lambda)v$. Since h is Hermitian, we have

$$\|\exp(-\lambda q)x\exp(\lambda q)\| \le \|x\|,$$

and this completes the proof.

Theorem 10. ([113].) Let $S \in B(X)$ be a scalar-type operator of class Γ.

(i) If F is any resolution of the identity for S of class Γ, it is of scalar type, i. e. $S = \int_{Sp(S)} \lambda\, F(d\lambda)$.

(ii) S has a unique resolution of the identity of class Γ.

Proof. (i) By [112] (or [147]) there is a quasi-nilpotent operator $N \in B(X)$ such that $S = R_0 + iJ_0 + N$ where

$$R_0 = \int_{Sp(S)} \operatorname{Re} \lambda\, F(d\lambda), \quad J_0 = \int_{Sp(S)} \operatorname{Im} \lambda\, F(d\lambda)$$

and N commutes with R_0 and J_0. By Theorem 4, we have $S = R + iJ$, where R, J are commuting Hermitian-equivalent. It follows from Lemma 9 that R_0, J_0 commute with R and J. By Theorem 4, R_0, J_0 are also Hermitian-equivalent. It follows from Lemma 8 that R, J, R_0, J_0 are all Hermitian with respect to an equivalent norm $|\cdot|_1$ and so N is normal with respect to $|\cdot|_1$. Thus $N = 0$ by NRI Theorem 5.14, and so $R = R_0$, $J = J_0$.

(ii) Let E be as in Theorem 4 and let F be of class Γ. It follows from (i) that

$$\int_{Sp(S)} p(\operatorname{Re} \lambda, \operatorname{Im} \lambda)\, E(d\lambda) = \int_{Sp(S)} p(\operatorname{Re} \lambda, \operatorname{Im} \lambda)\, F(d\lambda)$$

for any polynomial p of two variables, and hence

$$\int_{Sp(S)} \phi(\lambda)\, E(d\lambda) = \int_{Sp(S)} \phi(\lambda)\, F(d\lambda) \quad (\phi \in C(Sp(S))),$$

by the Stone-Weierstrass theorem. For $x \in X$, $f \in \Gamma$ the measures $f(E(\cdot)x)$, $f(F(\cdot)x)$ are countably additive and so regular. We have

$$\int_{Sp(S)} \phi(\lambda)\, f(E(d\lambda)x) = \int_{Sp(S)} \phi(\lambda)\, f(F(d\lambda)x) \quad (\phi \in C(Sp(S))),$$

and so, since $E(Sp(S)) = I$ for any resolution of the identity (Dunford [146], Theorem 1),

$$f(E(\cdot)x) = f(F(\cdot)x) \qquad (x \in X,\ f \in \Gamma).$$

Since Γ is total this gives $E = F$.

Dowson [144] has obtained appropriate extensions of Theorem 10 to arbitrary prespectral operators.

The final result shows that in order to obtain a satisfactory theory for scalar-type operators it is necessary to include condition (v) of the definition of a spectral measure.

Theorem 11. (Lumer [42].) Let E be a Boolean algebra of projections on X and let A be the closed real linear span of E. If the adjoint of each $T \in A$ is scalar-type of class X then E is uniformly bounded.

Proof. Let $T \in A$. Then T has real spectrum and so it follows from Theorem 4 that T^* is Hermitian-equivalent. Then T is Hermitian-equivalent and so E is Hermitian-equivalent by Lemma 8. A non-zero Hermitian projection has norm 1 by Sinclair's theorem and so we deduce that E is bounded.

7 · Further ranges

34. ESSENTIAL NUMERICAL RANGES

This section leans heavily on the lecture by J. P. Williams at the Conference on Numerical Ranges, Aberdeen, 1971. See Anderson [100], Anderson and Stampfli [101], Fillmore, Stampfli and Williams [151], Stampfli and Williams [67]. The main results will concern operators on Hilbert spaces, but the concepts can be introduced without extra difficulty on Banach spaces.

Let X denote a complex Banach space of infinite dimension, B the Banach algebra $B(X)$, $K = K(X)$ the set of all compact linear operators on X. Then K is a closed two-sided ideal of B, and we denote by π the canonical mapping of B onto the Banach algebra B/K.

Definition 1. Let $T \in B$. The <u>essential numerical range</u> $\mathrm{Vess}(T)$ of T is defined by

$$\mathrm{Vess}(T) = V(B/K, \pi(T)),$$

i. e. $\mathrm{Vess}(T)$ is the algebra numerical range of the canonical image $\pi(T)$ as an element of the unital Banach algebra B/K.

The following theorem lists some simple properties of $\mathrm{Vess}(T)$.

Theorem 2. <u>Let</u> $T \in B$.

(i) $\mathrm{Vess}(T)$ <u>is a non-void compact convex set.</u>

(ii) $\mathrm{Vess}(T) = \{0\}$ <u>if and only if</u> T <u>is compact.</u>

(iii) $\mathrm{Vess}(T) = \cap\{V(B, T+C) : C \in K\}$.

(iv) $\mathrm{Vess}(T) = \{f(T) : f \in D(I)$ <u>and</u> $f(K) = \{0\}\}$.

Proof. (i) NRI Theorem 2.3.

(ii) By NRI Theorem 4.1, $\mathrm{Vess}(T) = \{0\}$ if and only if

$\pi(T) = 0$, i.e. if and only if $T \in K$.

(iii) By Lemma 22.3, $V(B/K, \pi(T)) = \cap\{V(B, T+C) : C \in K\}$.

(iv) Let $\lambda \in Vess(T)$. Since B/K is a unital Banach algebra with unit element $1 = \pi(I)$, there exists $\phi \in D(B/K, 1)$ with $\phi(\pi(T)) = \lambda$. Let $f = \phi \circ \pi$. Then $f \in D(I)$, $f(K) = \phi(\pi(K)) = \{0\}$, and $f(T) = \lambda$.

On the other hand, given $f \in D(I)$ with $f(K) = \{0\}$, we have

$$f(T) = f(T + C) \in V(B, T + C) \qquad (C \in K),$$

and so $f(T) \in Vess(T)$, by (iii).

Definition 3. Let $T \in B$. The <u>Weyl spectrum</u> $WSp(T)$ of T is defined by

$$WSp(T) = \cap\{Sp(T + C) : C \in K\}.$$

The following elementary theorem is known as Weyl's theorem.

Theorem 4. *Let $T \in B$ and let $\lambda \in Sp(T)\backslash WSp(T)$. Then λ is an eigenvalue of T and $\lambda I - T$ has closed range and finite dimensional null space.*

Proof. There exists $C \in K$ such that $\lambda \notin Sp(T + C)$. Therefore $\lambda I - T - C$ is invertible. Let $A = (\lambda I - T - C)^{-1}C$. Then $A \in K$, and

$$\lambda I - T = (\lambda I - T - C)(I + A).$$

Since A is compact, $I + A$ has closed range and finite dimensional null space. Since $\lambda I - T - C$ is a linear homeomorphism of X onto X, the same is therefore true of $\lambda I - T$. Finally the null space of $I + A$ is non-zero since otherwise $I + A$ would be invertible, and then $\lambda I - T$ would be invertible.

Theorem 5. *Let $T \in B$. Then $coWSp(T) \subset Vess(T)$.*

Proof. Theorem 2 (i), (iii) and the inclusion $Sp(T + C) \subset V(B, T+C)$.

Remarks. (1) Theorems 4 and 5 show that each point of $Sp(T)\backslash Vess(T)$ is an eigenvalue of T with finite dimensional eigenspace.

(2) A brief discussion of operators T with $\mathrm{Vess}(T) \subset \mathbb{R}$ is given in Bonsall [120].

(3) $\mathrm{Vess}(T)$ has been defined in terms of the algebra numerical range. We could also define an essential numerical range in terms of the spatial numerical ranges of the operators $T + C$,

$$\mathrm{Vspess}(T) = \cap \{(V(T + C))^{-} : C \in K\}.$$

By the Zenger-Crabb theorem (Theorem 19. 4), we have $\mathrm{coSp}(T + C) \subset (V(T + C))^{-}$, and so

$$\mathrm{coWSp}(T) \subset \mathrm{Vspess}(T).$$

When X is a Hilbert space, $\mathrm{Vess}(T)$ and $\mathrm{Vspess}(T)$ coincide, because the closure of the spatial numerical range of an operator is then its closed convex hull.

Notation. For the rest of this section H will denote an infinite dimensional separable Hilbert space. In order to adhere to the established notation, for $T \in B(H)$ we shall denote by $W(T)$ the spatial numerical range of T, and by $\mathrm{Wess}(T)$ the essential numerical range $\mathrm{Vess}(T)$. We take $B = B(H)$, $K = K(H)$.

Given a closed linear subspace M of H, P_M will denote the (orthogonal) projection onto M, and $C_M(T)$ will denote the <u>compression</u> to M of an operator $T \in B(H)$, i. e.

$$C_M(T) = P_M T|_M.$$

Theorem 6. <u>Let</u> $T \in B$. <u>Then</u>

$$\mathrm{Wess}(T) = \cap\{V(B, T{+}C) : C \in K\} = \cap\{(W(T{+}C))^{-} : C \in K\}.$$

Proof. Since $W(T + C)$ is convex, $(W(T + C))^{-} = V(B, T + C)$.

Lemma 7. ([151].) <u>Let</u> M <u>be a closed linear subspace of</u> H <u>such that</u> $M^{\perp}$ <u>has finite dimension.</u> <u>Then</u>

$$\mathrm{Wess}(T) = \mathrm{Wess}(P_M T P_M) = \mathrm{Wess}(C_M(T)).$$

Proof. Let $P = P_M$. Then $I - P$ has finite rank, and so

$$T - PTP = (I - P)T(I - P) + PT(I - P) + (I - P)TP \in K.$$

Therefore $\pi(T) = \pi(PTP)$ (π the canonical mapping of B onto B/K), $\mathrm{Wess}(T) = \mathrm{Wess}(PTP)$.

Let I_M denote the identity operator on M. Given $f \in D(I_M)$ with $f(K(M)) = \{0\}$, define g on B by

$$g(A) = f(C_M(A)) \qquad (A \in B).$$

Since $C_M(I) = I_M$ and compression reduces norms, $g \in D(I)$. Also since the compression of a compact operator is compact, $g(K) = \{0\}$. Therefore, by Theorem 2 (iv),

$$\mathrm{Wess}(C_M(T)) \subset \mathrm{Wess}(T).$$

On the other hand, given $f \in D(I)$ with $f(K) = \{0\}$, let g be defined on $B(M)$ by

$$g(A) = f(AP) \qquad (A \in B(M)).$$

Since $I - P \in K$, we have $f(P) = f(I) = 1$, and so

$$g(I_M) = 1.$$

Thus $g \in D(I_M)$, and $g(K(M)) = \{0\}$, since $AP \in K$ whenever $A \in K(M)$. Therefore, by Theorem 2 (iv) again,

$$f(PTP) = f(C_M(T)P) = g(C_M(T)) \in \mathrm{Wess}(C_M(T)),$$

and so $\mathrm{Wess}\,(PTP) \subset \mathrm{Wess}(C_M(T))$.

Lemma 8. ([101].) Let $\lambda \in \mathrm{Wess}(T)$. Then there exist a closed linear subspace E of H with infinite dimension, an orthonormal basis $\{e_k\}$ for E, and complex numbers λ_k such that

(i) $\lim_{k\to\infty} \lambda_k = \lambda$,

(ii) $C_E(T)$ has the matrix representation $\mathrm{diag}\{\lambda_k\}$ relative to the basis $\{e_k\}$, i.e.

$$(Te_j, e_i) = \begin{cases} \lambda_i & (j = i) \\ 0 & (j \neq i) . \end{cases}$$

Proof. By Theorem 6, $\lambda \in (W(T))^-$. Therefore there exists a unit vector $e_1 \in H$ with $|(Te_1, e_1) - \lambda| < 1$.

Suppose that an orthonormal n-tuple $(e_1, \ldots, e_n)$ has been chosen in H such that

$$(Te_j, e_i) = 0 \quad (i \neq j), \qquad |(Te_i, e_i) - \lambda| < \frac{1}{i}$$

for $i, j = 1, 2, \ldots, n$. Take M to be the orthogonal complement of the linear span of the $3n$ vectors $e_1, \ldots, e_n, Te_1, \ldots, Te_n, T^*e_1, \ldots, T^*e_n$. By Lemma 7, $\lambda \in \mathrm{Wess}(C_M(T))$. Therefore, by the first step in the proof applied to $C_M(T)$ in place of T, there exists a unit vector e_{n+1} in M such that

$$|(C_M(T)e_{n+1}, e_{n+1}) - \lambda| < \frac{1}{n+1} .$$

Then, since $e_{n+1} \in M$, $(e_1, \ldots, e_{n+1})$ is an orthonormal (n+1)-tuple, $(Te_j, e_i) = 0$ $(i, j = 1, \ldots, n+1, \ i \neq j)$, and $|(Te_{n+1}, e_{n+1}) - \lambda| < \frac{1}{n+1}$. In this way we obtain a countable orthonormal set $\{e_k\}$ with the required properties, and we take E to be the closed linear hull of this set.

Remark. By taking E to be the closed linear hull of $\{e_{2k} : k = 1, 2, \ldots\}$ we can arrange that both E and $E^\perp$ have infinite dimension, without other change in the lemma.

Theorem 9. ([101], [151].) <u>Let</u> $T \in B$. <u>Then the following statements are equivalent.</u>

(1) <u>There exists a closed linear subspace</u> M <u>of</u> H <u>with infinite dimension such that</u> $C_M(T)$ <u>has the matrix representation</u> $\mathrm{diag}\{\lambda_k\}$ <u>relative to an orthonormal basis for</u> M, <u>and</u> $\lim_{k\to\infty} \lambda_k = \lambda$.

(2) <u>There exists a projection</u> $P \in B$ <u>of infinite rank such that</u> $P(T - \lambda I)P$ <u>is compact.</u>

(3) <u>There exists a countable orthonormal set</u> $\{e_k : k = 1, 2, \ldots\}$ <u>such that</u>

$$\lim_{k\to\infty}(Te_k, e_k) = \lambda.$$

(4) <u>There exists a sequence</u> $\{x_k\}$ <u>of unit vectors in</u> H <u>that converges weakly to zero and satisfies</u>

$$\lim_{k\to\infty} (Tx_k, x_k) = \lambda.$$

(5) $\lambda \in \text{Wess}(T)$.

Proof. (1) $\Rightarrow$ (2). Let M be a closed linear subspace of H with infinite dimension such that $C_M(T)$ has the matrix representation $\text{diag}\{\lambda_k\}$ relative to an orthonormal basis $\{e_k\}$ for M, and $\lim_{k\to\infty} \lambda_k = \lambda$. Let P be the projection onto M, and let D be the unique bounded linear operator on M such that

$$De_k = (\lambda_k - \lambda)e_k \qquad (k = 1, 2, \ldots).$$

Since $\lim_{k\to\infty} (\lambda_k - \lambda) = 0$, $D \in K(M)$. By hypothesis,

$$((C_M(T) - \lambda I_M)e_j, e_i) = (De_j, e_i) \qquad (i, j = 1, 2, \ldots),$$

and so

$$C_M(T) - \lambda I_M = D.$$

Therefore

$$PTP - \lambda P = DP \in K.$$

(2) $\Rightarrow$ (3). Let P be a projection belonging to B of infinite rank such that $PTP - \lambda P \in K$, and let $\{u_n\}$ be an orthonormal basis for PH. Then there exists a subsequence $\{u_{n_k}\}$ and $x \in PH$ such that $\lim_{k\to\infty} (PTP - \lambda P)u_{n_k} = x$. Take $e_k = u_{n_k}$ $(k = 1, 2, \ldots)$. We have

$$(PTPe_k, e_k)-\lambda = ((PTP-\lambda P)e_k, e_k) = (x, e_k) + ((PTP-\lambda P)e_k-x, e_k).$$

Since $\{e_k\}$ is orthonormal, $\lim_{k\to\infty} (x, e_k) = 0$, and so $\lim_{k\to\infty} (PTPe_k, e_k)=\lambda$. Since $Pe_k = e_k$, we have

$$(PTPe_k, e_k) = (TPe_k, Pe_k) = (Te_k, e_k),$$

and (3) is proved.

(3) $\Rightarrow$ (4). An orthonormal sequence converges weakly to zero.

(4) $\Rightarrow$ (5). Let $\{x_k\}$ be a sequence of vectors with $\|x_k\| = 1$ that converges weakly to zero and satisfies $\lim_{k\to\infty} (Tx_k, x_k) = \lambda$. For each $C \in K$, we have $\lim_{k\to\infty} \|Cx_k\| = 0$, and so

$$\lim_{k\to\infty} ((T + C)x_k, x_k) = \lambda.$$

Therefore $\lambda \in (W(T + C))^-$ $(C \in K)$, and Theorem 6 gives $\lambda \in \text{Wess}(T)$.

(5) $\Rightarrow$ (1). Lemma 8.

Corollary 10. There exists a compact compression of T onto an infinite dimensional closed linear subspace of H if and only if $0 \in \text{Wess}(T)$.

Proof. Let M be an infinite dimensional closed linear subspace and let $P = P_M$. Then $PTP = C_M(T)P$ and $C_M(T) = PTP|_M$. Thus $C_M(T)$ is compact if and only if PTP is compact. The equivalence of (2) and (5) (with $\lambda = 0$) completes the proof.

Remark. While most of the arguments used here are available only in a Hilbert space context, the implication (4) $\Rightarrow$ (5) is available for Banach spaces in the following form.

Let $T \in B(X)$, $(x_k, f_k) \in \Pi(X)$, $\lim_{k\to\infty} f_k(Tx_k) = \lambda$, and let $\{x_k\}$ converge weakly to zero. Then $\lambda \in \text{Vess}(T)$, (in fact $\lambda \in \text{Vspess}(T)$).

Theorem 9 is the starting point for the proof of the following remarkable theorem of Anderson [100]. We omit the proof as it would take us too deeply into Hilbert space theory.

Theorem 11. Let $T \in B$. Then $0 \in \text{Wess}(T)$ if and only if $T = AX - XA$ for some $A, X \in B$ with $A^* = A$.

Theorem 11 includes as a corollary the characterization due to Radjavi [195] of self-adjoint operators of the form

X*X - XX* as the self-adjoint operators A for which $0 \in \text{Wess}(A)$. It has also been used by Anderson [100] together with an argument from Williams [221] involving similarity in Banach algebras to prove the following sharper form of a theorem of Brown and Pearcy [127] characterizing commutators.

Theorem 12. Let $T \in B$. Then $T = XY - YX$ for some $X, Y \in B$ if and only if, for every non-zero λ, $T - \lambda I \notin K$. If this condition is satisfied then X can be chosen to be similar to a self-adjoint operator.

35. JOINT NUMERICAL RANGES

We now generalize some of the results of §§23, 24 to joint algebra numerical ranges. The methods employed are essentially similar to the one variable case but the details are more cumbersome. For most of this section we shall simply state the results, supplying the details only for illustrative examples. The construction of the extremal algebras was foreshadowed in §25.

We also include in this section Dekker's extension [138] of the Toeplitz-Hausdorff theorem for joint spatial numerical ranges in Hilbert space.

Let A be a complex unital Banach algebra and let $a_1, a_2, \ldots, a_n \in A$. We recall (NRI Definition 2.11) that the joint numerical range of $a_1, a_2, \ldots, a_n$ is defined by

$$V(a_1, a_2, \ldots, a_n) = \{(f(a_1), f(a_2), \ldots, f(a_n)) : f \in D(1)\}$$

and is a compact convex subset of $\underset{\sim}{C}^n$. For simplicity of exposition we shall consider the case $n = 2$. Let Ξ be a compact convex subset of $\underset{\sim}{C}^2$, and let

$$\omega(\lambda, \mu) = \sup\{|\exp(\lambda z + \mu w)| : (z, w) \in \Xi\} \quad (\lambda, \mu \in \underset{\sim}{C}).$$

We easily obtain the following analogue of Lemma 24.6.

Lemma 1. Given $a, b \in A$, we have $V(a, b) \subset \Xi$ if and only if

$$\|\exp(\lambda a + \mu b)\| \le \omega(\lambda, \mu) \qquad (\lambda, \mu \in \underset{\sim}{C}).$$

Lemma 1 leads to estimates for the norms of polynomials in a and b, when a, b commute.

Lemma 2. Let $a, b \in A$, $ab = ba$, $V(a, b) \subset \Xi$, and let r, s be non-negative integers. Then

$$\|a^r b^s\| \le r!\,s! \inf_{\alpha, \beta > 0} \alpha^{-r}\beta^{-s} \sup\{\omega(\lambda, \mu) : |\lambda| = \alpha,\ |\mu| = \beta\}.$$

Proof. Use the formula

$$a^r b^s = (2\pi i)^{-2} r!\,s! \int_{\Gamma_\alpha} \int_{\Gamma_\beta} \lambda^{-(r+1)} \mu^{-(s+1)} \exp(\lambda a + \mu b)\, d\lambda\, d\mu.$$

Example 3. Let $\Xi = \{(z, w) : |z| + |w| \le 1\}$ in Lemma 2. Then

$$\|a^r b^s\| \le r!\,s! \left(\frac{e}{r+s}\right)^{r+s}$$

Proof. For this case we have

$$\begin{aligned} \omega(\lambda, \mu) &= \sup\{|\exp(\lambda z + \mu w)| : |z| + |w| = 1\} \\ &= \exp(\max\{|\lambda|, |\mu|\}), \end{aligned}$$

and hence, by taking $|\lambda| = |\mu| = t$,

$$\|a^r b^s\| \le r!\,s! \inf_{t > 0} t^{-r-s} e^t = r!\,s! \left(\frac{e}{r+s}\right)^{r+s}$$

Note that this estimate is much sharper than the product of the one variable estimates given by NRI Theorem 4. 8.

The extremal algebras are constructed as in §24. In detail, let $M^\omega(\underset{\sim}{C}^2)$ be the Banach space of all (finite) complex regular Borel measures ν on $\underset{\sim}{C}^2$ where

$$\|\nu\|_\omega = \int \omega\, d|\nu|\,.$$

Given $\nu \in M^{\omega}(\underset{\sim}{C}^2)$ we define f_ν on Ξ by

$$f_\nu(z, w) = \int e^{z\lambda + w\mu} \, d\nu(\lambda, \mu) \qquad ((z, w) \in \Xi).$$

We denote by $Ea(\Xi)$ the set of all functions f_ν with $\nu \in M^{\omega}(\underset{\sim}{C}^2)$ and we define, for $f \in Ea(\Xi)$,

$$\|f\| = \inf\{\|\nu\|_\omega : \nu \in M^{\omega}(\underset{\sim}{C}^2), \ f_\nu = f\}.$$

Then $Ea(\Xi)$ is a Banach algebra for which the analogue of Theorem 24.2 holds.

The dual of $Ea(\Xi)$ is again given by a space of entire functions. In detail, let $D(\Xi)$ be the set of all entire functions ϕ of two complex variables such that

$$\|\phi\| = \sup\{\frac{|\phi(\lambda, \mu)|}{\omega(\lambda, \mu)} : \lambda, \mu \in \underset{\sim}{C}\} < \infty .$$

Then $D(\Xi)$ is isometrically isomorphic to $Ea(\Xi)'$ under the mapping $\phi \to \Phi_\phi$ where

$$\Phi_\phi(f) = \int \phi \, d\nu \qquad (\nu \in M^{\omega}(\underset{\sim}{C}^2), \ f_\nu = f).$$

If now a, b are elements of a Banach algebra for which $ab = ba$ and $V(a, b) \subset \Xi$, then the analogue of Theorem 24.9 holds.

As an illustration of the above we show that the estimate in Example 3 is best possible.

Example 4. Let $\Xi = \{(z, w) : |z| + |w| \leq 1\}$, let $u(z, w) = z$, $v(z, w) = w$. Then, in the algebra $Ea(\Xi)$,

$$\|u^r v^s\| = r!\, s!\, (\frac{e}{r+s})^{r+s}$$

Proof. Let

$$\phi(\lambda, \mu) = (\frac{e}{r+s})^{r+s} \lambda^r \mu^s \qquad (\lambda, \mu \in \underset{\sim}{C})$$

and let $t = \max\{|\lambda|, |\mu|\}$. Then

$$|\phi(\lambda,\mu)| \le \left(\frac{et}{r+s}\right)^{r+s} \le e^t \quad (t \ge 0).$$

Thus $\phi \in D(\Xi)$, $\|\phi\| = 1$, and so

$$\|u^r v^s\| \ge r!\,s!\,\left(\frac{e}{r+s}\right)^{r+s}$$

Example 3 completes the proof.

Remark. For any Ξ invariant under rotations of the coordinates it is easily seen by the above arguments that

$$\|u^r v^s\| = r!\,s! \inf_{\sigma,\tau > 0} \sigma^{-r}\,\tau^{-s}\,\omega(\sigma,\tau).$$

When Ξ is a Cartesian product we can relate $Ea(\Xi)$ to one variable extremal algebras. Let K_1, K_2 be compact convex subsets of $\mathbb{C}$ and let $\Xi = K_1 \times K_2$. Then it may be shown that $Ea(\Xi)$ is the projective tensor product $Ea(K_1) \hat{\otimes} Ea(K_2)$.

Finally we consider the joint spatial numerical range for operators on a Hilbert space H. Given $T_1, T_2, \dots, T_n \in B(H)$ we define the joint spatial numerical range $W(T_1, T_2, \dots, T_n)$ by

$$W(T_1, T_2, \dots, T_n) = \{((T_1x, x), (T_2x, x), \dots, (T_nx, x)) : x \in H, \|x\| = 1\}.$$

Theorem 5. (Dekker [138].) <u>Let</u> $T_1, T_2, \dots, T_n$ <u>be commuting normal operators in</u> $B(H)$. <u>Then</u> $W(T_1, T_2, \dots, T_n)$ <u>is a convex subset of</u> $\mathbb{C}^n$.

Proof. By the spectral theorem for commuting normal operators (see for example Segal [200]) there is a unitary transformation $U : H \to L_2(\mu)$ for some measure μ such that

$$U^{-1}T_j U f = \phi_j f \quad (f \in L_2(\mu)) \quad (j = 1, 2, \dots, n),$$

where ϕ_j is bounded and μ-measurable. Since the joint spatial numerical range is invariant under unitary transformations we may suppose without loss that $H = L_2(\mu)$, $T_j f = \phi_j f$.

Let $f, g \in S(H)$, $0 \leq \alpha \leq 1$, and let

$$h = (\alpha|f|^2 + (1 - \alpha)|g|^2)^{\frac{1}{2}}.$$

Then $h \in S(H)$ and for $j = 1, 2, \ldots, n$,

$$\begin{aligned}(T_j h, h) &= \int \phi_j \; |h|^2 \, d\mu \\ &= \alpha \int \phi_j \; |f|^2 \, d\mu + (1 - \alpha) \int \phi_j \; |g|^2 \, d\mu \\ &= \alpha(T_j f, f) + (1 - \alpha)(T_j g, g).\end{aligned}$$

The result follows.

Remarks. (1) If $T_1, \ldots, T_n$ are self-adjoint operators, not necessarily commuting with each other, then $W(T_1, T_2, \ldots, T_n)$ is a convex subset of $\underset{\sim}{R}^n$. The proof is an obvious extension of the proof of Lemma 15.10.

(2) That $W(T_1, T_2)$ is not convex in general is easily seen from the following two-dimensional example. Let

$$T_1 = \begin{bmatrix} 1 & 0 \\ 0 & 0 \end{bmatrix}, \qquad T_2 = \begin{bmatrix} 0 & 0 \\ 1 & 0 \end{bmatrix}$$

and then we easily obtain

$$W(T_1, T_2) = \{(r^2, rse^{i\theta}) : r, s \geq 0, \; r^2 + s^2 = 1, \; \theta \in \underset{\sim}{R}\}.$$

It is clear that $(\alpha, 0) \in W(T_1, T_2)$ if and only if $\alpha = 0$ or $\alpha = 1$, and so $W(T_1, T_2)$ is not convex.

(3) Asplund and Pták [105] consider a related concept of a two-dimensional range for a pair of linear mappings on a Banach space.

36. MATRIX RANGES

It is natural to generalize the numerical ranges of operators by replacing sets of complex numbers by sets of $n \times n$ matrices. This has been carried out most notably by W. B. Arveson in an important series of papers [102, 103, 104]. Given $T \in B(H)$, H a Hilbert space, Arveson defines $\mathscr{W}_n(T)$ to be the set of all $\phi(T)$ obtained by letting ϕ run through the set of all normalized 'completely positive' mappings of B(H) into $B(\underset{\sim}{C}^n)$. When $n = 1$, these completely positive mappings coincide with the normalized states on B(H), and so $\mathscr{W}_1(T) = V(B(H), T)$ the algebra numerical range of T. Thus the matrix ranges $\mathscr{W}_n(T)$ generalize the algebra numerical range rather than the spatial numerical range (i. e. the classical numerical range W(T)). As might be expected the sets $\mathscr{W}_n(T)$ are highly regular, in fact they are compact and convex in a strong sense appropriate for matrices. Arveson demonstrates the power of the concept by proving that the set $\{\mathscr{W}_n(T) : n = 1, 2, \dots\}$ constitutes a complete set of unitary invariants for the irreducible compact linear operators T on H.

S. K. Parrott (private communication) has considered the concept, denoted here by $W_n(T)$, which generalizes in a most direct and natural way the classical numerical range W(T). These ranges had also been considered by Halmos [30, 156] who mapped them back into the complex plane by taking the traces of the matrices. Parrott shows that the $W_n(T)$ $(n = 1, 2, \dots)$ constitute a complete set of unitary invariants for the compact linear operators T on H with zero reducing null spaces. We content outselves mainly with the theory of the ranges $W_n(T)$ which are technically simpler to discuss than the $\mathscr{W}_n(T)$. Thus our account is mainly based on the work of Parrott. We also consider the relationship between $W_n(T)$ and $\mathscr{W}_n(T)$, showing that

$$\mathscr{W}_n(T) = W_n(\chi(T)),$$

where χ is a certain representation of B(H) which we call the 'great universal representation'. We use this representation to derive a few of the elementary properties of $\mathscr{W}_n(T)$.

So far as we are aware, matrix ranges of operators on Banach spaces have not yet been developed.

Notation. H will denote a Hilbert space with scalar product (,), $\underset{\sim}{N}$ the set of natural numbers. With $n \in \underset{\sim}{N}$, $\underset{\sim}{C}^n$ will be regarded as a Hilbert space with the usual scalar product

$$(\underset{\sim}{\xi}, \underset{\sim}{\eta}) = \sum_{k=1}^{n} \xi_k \eta_k^* \quad (\underset{\sim}{\xi} = \{\xi_k\},\ \underset{\sim}{\eta} = \{\eta_k\} \in \underset{\sim}{C}^n).$$

$\underset{\sim}{O}_n$ denotes the set of all orthonormal n-tuples $\underset{\sim}{u} = (u_1, \ldots, u_n)$ of vectors in H.

$\underset{\sim}{P}_n$ denotes the set of all (orthogonal) projections on H of rank n, i.e. $\dim(PH) = n$.

$\underset{\sim}{Q}_n$ denotes the set of all linear isometries of $\underset{\sim}{C}^n$ into H.

I_n denotes the identity operator on $\underset{\sim}{C}^n$.

Definition 1. Let $T \in B(H)$. The nth <u>spatial matrix range</u> is the set $W_n(T)$ of linear operators on $\underset{\sim}{C}^n$ given by

$$W_n(T) = \{V^*TV : V \in \underset{\sim}{Q}_n\}.$$

We have chosen this among several possible definitions for $W_n(T)$ because of its simplicity and convenience in use. The following lemma gives a highly intuitive expression for $W_n(T)$ in terms of matrices and exhibits $W_n(T)$ as a generalization of the classical numerical range $W(T)$.

Lemma 2. <u>Given</u> $\underset{\sim}{u} = (u_1, u_2, \ldots, u_n) \in \underset{\sim}{O}_n$, <u>let</u> $A(\underset{\sim}{u}, T)$ <u>denote the</u> $n \times n$ <u>matrix</u> (α_{ij}) <u>with</u> $\alpha_{ij} = (Tu_j, u_i)$ $(i, j = 1, 2, \ldots, n)$. <u>Then</u>

$$\{A(\underset{\sim}{u}, T) : \underset{\sim}{u} \in \underset{\sim}{O}_n\}$$

<u>coincides with the set of matrix representations relative to the natural basis for</u> $\underset{\sim}{C}^n$ <u>of the operators in</u> $W_n(T)$.

Proof. Let $e_1, \ldots, e_n$ be the natural basis for $\underset{\sim}{C}^n$, i.e. e_k is the n-tuple with 1 in the kth place and 0 elsewhere. Given $V \in \underset{\sim}{Q}_n$, we have

$$\underset{\sim}{u} = (Ve_1, \ldots, Ve_n) \in \underset{\sim}{O}_n,$$

and, conversely, each $\underset{\sim}{u} \in \underset{\sim}{O}_n$ is of this form for some $V \in \underset{\sim}{Q}_n$. With V and $\underset{\sim}{u}$ related in this way, we have

$$(V^*TVe_j, e_i) = (TVe_j, Ve_i) = (Tu_j, u_i);$$

i.e. $A(\underset{\sim}{u}, T)$ is the matrix representing V^*TV relative to the basis

$e_1, \ldots, e_n$.

$W_n(T)$ can also be obtained in terms of $\underset{\sim}{P}_n$, as the following lemma shows. This is the form in which the concept was introduced by Halmos [30, 156] and S. K. Parrott.

Lemma 3. (i) $\underset{\sim}{P}_n = \{VV^* : V \in \underset{\sim}{Q}_n\}$,

(ii) The operators of the form $PT|_{PH}$ with $P \in \underset{\sim}{P}_n$ are unitarily equivalent to the operators in $W_n(T)$.

Proof. (i) Given $V \in \underset{\sim}{Q}_n$, $V^*V = I_n$ (the identity operator on $\underset{\sim}{C}_n$). Thus

$$(VV^*)^2 = VI_nV^* = VV^*,$$

and

$$V\underset{\sim}{C}^n \supset VV^*H \supset VV^*V\underset{\sim}{C}^n = VI_n\underset{\sim}{C}^n = V\underset{\sim}{C}^n,$$

which together show that $VV^* \in \underset{\sim}{P}_n$. Also, given $P \in \underset{\sim}{P}_n$, there exists an isometric linear mapping V of $\underset{\sim}{C}^n$ onto PH, and then $VV^* = P$.

(ii) Let $V \in \underset{\sim}{Q}_n$ and $P = VV^*$. Regarded as a mapping of $\underset{\sim}{C}^n$ onto PH, V is a unitary mapping, U say, and we have

$$U^* \, PT|_{PH} \, U = V^*VV^*TV = V^*TV.$$

Lemma 4. Let $S, T \in B(H)$ and $W_n(T) \subset W_n(S)$. Then, for each $P \in \underset{\sim}{P}_n$, there exists a partial isometry G with initial projection P such that

$$PTP = G^*SG.$$

Proof. By Lemma 3(i), given $P \in \underset{\sim}{P}_n$, we have $P = UU^*$ for some $U \in \underset{\sim}{Q}_n$. Then

$$U^*TU \in W_n(T) \subset W_n(S).$$

Therefore there exists $V \in \underset{\sim}{Q}_n$ with

$$U^*TU = V^*SV.$$

Take $G = VU^*$. Then

$$G^*G = UV^*VU^* = UI_nU^* = UU^* = P;$$

which shows that G is a partial isometry with initial projection P. Also

$$PTP = UU^*TUU^* = U(V^*SV)U^* = G^*SG.$$

Remarks. (1) It follows at once from the proof of Lemma 4 that, given $P \in \underset{\sim}{P}_n$, there exists $A \in W_n(S)$ with

$$\|A\| = \|PTP\|.$$

For, in the notation of that proof, $V^*SV = A \in W_n(S)$, and

$$\|PTP\| = \|UAU^*\| = \|AU^*\| = \|UA^*\| = \|A^*\| = \|A\|.$$

(2) If $W_n(T) \subset W_n(S)$, we also have

$$W_n(\text{Re } T) \subset W_n(\text{Re } S), \quad W_n(\text{Im } T) \subset W_n(\text{Im } S).$$

To see this, let $S = A + iB$, $T = C + iD$ with A, B, C, D self-adjoint. Given $V \in \underset{\sim}{Q}_n$, there exists $U \in \underset{\sim}{Q}_n$ such that

$$V^*TV = U^*SU,$$

i.e. $\quad V^*CV + iV^*DV = U^*AU + iU^*BU.$

Since V^*CV, V^*DV, U^*AU, U^*BU are all self-adjoint, this gives

$$V^*CV = U^*AU, \qquad V^*DV = U^*BU,$$

and so $W_n(C) \subset W_n(A)$, $W_n(D) \subset W_n(B)$.

(3) It is proved in Halmos [30, Problem 167] that $\text{tr } W_n(T)$ is convex (the first proof being attributed to C. A. Berger). That $W_n(T)$ is not itself convex in general is easy to see. Let $H = \underset{\sim}{C}^n$. Then $\underset{\sim}{Q}_n$ is the set of unitary mappings of $\underset{\sim}{C}^n$ onto $\underset{\sim}{C}^n$, and so $W_n(T)$ is the set of unitary equivalents of T. In particular, let T be itself unitary.

Then $W_n(T)$ is the conjugacy class of T in the unitary group. In this case $W_n(T)$ is convex if and only if $T = e^{i\theta}I$ for some $\theta \in \mathbb{R}$. For if $U, V \in W_n(T)$ and $\frac{1}{2}(U + V) \in W_n(T)$, then we have in turn
$\frac{1}{4}(U + V)^*(U + V) = I$, $U^*V + VU^* = 2I$, $(U - V)^*(U - V) = 0$, $U - V = 0$.
Thus we have $V^*TV = T$ for all unitary V, T commutes with all elements of $B(H)$, $T = e^{i\theta}I$ for some $\theta \in \mathbb{R}$.

Notation. Let τ_w, τ_s, τ_n denote the weak operator, strong operator, and norm topologies on $B(H)$, and let $B_1(H)$ denote the closed unit ball in $B(H)$.

The following elementary lemma is well known.

Lemma 5. Multiplication of operators is jointly continuous as a mapping:

(i) $(B_1(H), \tau_w) \times (B(H), \tau_s) \to (B(H), \tau_w)$,

(ii) $(B_1(H), \tau_s) \times (B(H), \tau_s) \to (B(H), \tau_s)$.

Proof. With $A, A_0 \in B_1(H)$, $B, B_0 \in B(H)$, $x, y \in H$, we have

$$\begin{aligned} |((AB - A_0B_0)x, y)| &\le |(A(B - B_0)x, y)| + |((A - A_0)B_0x, y)| \\ &\le \|(B - B_0)x\| . \|y\| + |((A - A_0)B_0x, y)|. \end{aligned}$$

This proves (i), and (ii) is even easier.

Theorem 6. Let $S, T \in B(H)$ and $W_n(T) \subset W_n(S)$ $(n = 1, 2, \ldots)$. Then

$$\|T\| \le \|S\|.$$

Proof. Let Λ denote the directed set of all finite dimensional subspaces of H, and for each $E \in \Lambda$, let P_E denote the projection onto E. Then $P_E \to I$ (τ_s) along the directed set Λ, for given any finite subset F of H, we have $P_Ex = x$ $(x \in F)$ whenever $E \supset F$. By Lemma 5(ii), it follows that $P_ETP_E \to T$ (τ_s) along Λ. By Lemma 4, there exists a partial isometry G_E with initial projection P_E such that

$$P_E T P_E = G_E^* S G_E,$$

and so $\|P_E T P_E\| \le \|S\|$. The result follows.

Theorem 7. Let $S, T \in B(H)$, let S be compact, and let

$$W_n(T) \subset W_n(S) \qquad (n = 1, 2, \ldots).$$

Then T is compact, and there exists $G \in B_1(H)$ such that

$$T = G^* S G.$$

Proof. Suppose first that P is the projection onto an arbitrary closed separable linear subspace of H. Then there exist projections P_n of finite rank $(n = 1, 2, \ldots)$ such that $P = \lim_{n\to\infty} P_n\ (\tau_s)$, and by Lemma 4 there exist partial isometries G_n with initial projections P_n such that

$$P_n T P_n = G_n^* S G_n \qquad (n = 1, 2, \ldots).$$

Let H_0 denote the closed linear span of $\{PH, P_nH, G_nH\ (n=1, 2, \ldots)\}$, and note that H_0 contains the subspaces G_n^*H, since the final space of G_n^* is the initial space of G_n. Let $F_n = G_n|_{H_0}$. Then F_n is a partial isometry with initial space P_nH.

Since H_0 is separable, $B_1(H_0)$ with the topology τ_w is metrizable and compact (see Dixmier [139, p. 32]). Therefore $\{F_n\}$ has a τ_w-convergent subsequence, and throwing away the rest of the sequence, we may suppose that there exists $F \in B_1(H_0)$ such that $F = \lim_{n\to\infty} F_n\ (\tau_w)$. Define G on H by taking

$$G(x + y) = Fx \qquad (x \in H_0,\ y \in H_0^{\perp}).$$

Clearly $G \in B_1(H)$; and, since G_n and G_n^* vanish on $H_0^{\perp}$, we have

$$(G_n(x+y),\ x' + y') = (F_n x, x') \quad (x, x' \in H_0,\ y, y' \in H_0^{\perp}),$$

from which we have $G = \lim_{n\to\infty} G_n(\tau_w)$.

Since S is compact, we now have

$$SG = \lim_{n\to\infty} SG_n \quad (\tau_s).$$

Also $G^* = \lim_{n\to\infty} G_n^* \ (\tau_w)$, and $G^*, G_n^* \in B_1(H)$. Therefore, by Lemma 5(i),

$$G^*SG = \lim_{n\to\infty} G_n^*SG_n \qquad (\tau_w).$$

But we also have $PTP = \lim_{n\to\infty} P_nTP_n \ (\tau_s)$, and therefore

$$PTP = G^*SG.$$

We have now proved that PTP is compact for each projection P onto a closed separable linear subspace of H. To prove that T is compact, let $x_n \in H$, $\|x_n\| \le 1$ $(n = 1, 2, \ldots)$, and take P to be the projection onto the closed linear span of the set $\{x_1, x_2, \ldots, Tx_1, Tx_2, \ldots\}$. Then PTP is compact and so there exists a subsequence $\{x_{n_k}\}$ such that $\{PTPx_{n_k}\}$ converges. But, by definition of P,

$$PTPx_{n_k} = Tx_{n_k},$$

and so T is compact.

Since T is compact, the subspace $(TH + T^*H)^-$ is separable, and we now take P to be the projection onto this subspace. Then $PT = T$ and $PT^* = T^*$, from which $TP = T$. Therefore

$$T = PTP = G^*SG.$$

Remark. The proof of Theorem 7 provides some more detailed information about G, as follows. Let P be the projection onto $(TH + T^*H)^-$. Then there exist projections P_n of finite rank and partial isometries G_n with initial projections P_n, such that

$$P = \lim_{n\to\infty} P_n \ (\tau_s), \quad G = \lim_{n\to\infty} G_n \ (\tau_w), \quad P_nTP_n = G_n^*SG_n \quad (n=1, 2, \ldots),$$

$$G^*SG = T = \lim_{n\to\infty} P_nTP_n = \lim_{n\to\infty} P_nT = \lim_{n\to\infty} TP_n \quad (\tau_n).$$

Everything here has been proved except the last two equalities and that this convergence is with respect to the norm topology. To see this, note that since T is compact, $P = \lim_{n\to\infty} P_n \ (\tau_s)$, and $\{P_n\}$ is an equicontinuous family, we have

$$T = PT = \lim_{n\to\infty} P_n T \qquad (\tau_n),$$

and similarly

$$T^* = \lim_{n\to\infty} P_n T^* \qquad (\tau_n).$$

It is easy to see that if S and T are unitarily equivalent, then $W_n(S) = W_n(T)$. For if U is a unitary operator on H with $T = U^*SU$, then

$$\begin{aligned} W_n(T) = \{V^*TV : V \in \underset{\sim}{Q}_n\} &= \{V^*U^*SUV : V \in \underset{\sim}{Q}_n\} \\ &= \{(UV)^*SUV : V \in \underset{\sim}{Q}_n\} = W_n(S). \end{aligned}$$

We come now to the main theorem of this section (due to S. K. Parrott), showing that the spatial matrix ranges $W_n(T)$ for $n = 1, 2, \ldots$ constitute a complete set of unitary invariants for compact operators T that have zero reducing null-spaces.

Definition 8. Given $T \in B(H)$, the reducing null-space of T is the intersection of the kernels of T and T^*, i. e.

$$\{x \in H : Tx = T^*x = 0\}.$$

Theorem 9. Let S, T be compact linear operators on H with zero reducing null-spaces, and suppose that

$$W_n(S) = W_n(T) \qquad (n = 1, 2, \ldots).$$

Then S, T are unitarily equivalent.

Proof. By Theorem 7, there exist $F, G \in B_1(H)$ with

$$S = F^*TF, \qquad T = G^*SG.$$

Let $Q = FG$. Then $Q \in B_1(H)$ and

$$T = Q^*TQ. \tag{1}$$

We have $T^*T = Q^*(T^*QQ^*T)Q$, and so, taking $A = T^*T$, $B = (Q^*T)^*Q^*T$, we have

$$A = Q^*BQ, \qquad \|Q\| \leq 1, \qquad A \geq B \geq 0\,, \tag{2}$$

for $(Ax, x) - (Bx, x) = \|Tx\|^2 - \|Q^*Tx\|^2 \geq 0$.

In the following argument, which is due essentially to Douglas [142], we make frequent use of the elementary fact that

$$L, M \in B(H), \qquad M \geq 0, \qquad L^*ML = 0 \Rightarrow ML = 0, \tag{3}$$

which follows at once from $\|M^{\frac{1}{2}}L\|^2 = \|(M^{\frac{1}{2}}L)^*M^{\frac{1}{2}}L\| = \|L^*ML\| = 0$.

We prove that

$$Q^*QA = QQ^*A = A. \tag{4}$$

Assume that $A \neq 0$, and let E denote the projection onto the eigenspace corresponding to the eigenvalue $\|A\|$ of the positive compact operator A. We have $EA = AE = \|A\|E$, and so

$$\begin{aligned}\|A\|E = EAE &= (QE)^*B(QE)\\ &\leq (QE)^*A(QE) \leq \|A\|(QE)^*(QE) \leq \|A\|E.\end{aligned} \tag{5}$$

Equality holds throughout (5), and therefore

$$E(I - Q^*Q)E = E - (QE)^*QE = 0, \tag{6}$$

and

$$(QE)^*(\|A\|I - A)(QE) = 0. \tag{7}$$

We have $I - Q^*Q \geq 0$ and $\|A\|I - A \geq 0$, and so the remark (3) applied to (6) and (7) gives

$$(I - Q^*Q)E = 0, \tag{8}$$

$$(\|A\|I - A)QE = 0. \tag{9}$$

By (9), QEH is contained in the eigenspace EH, i.e. Q maps EH into EH. Since EH has finite dimension, (8) now shows that Q maps EH isometrically onto EH and that Q^* is the inverse mapping. Thus EH is a reducing subspace for Q and

$$Q^*QE = QQ^*E = E. \tag{10}$$

From the equality everywhere in (5) we also have

$$(QE)^*(\|A\|I - B)(QE) = 0,$$

and another application of (3) gives

$$(\|A\|I - B)QE = 0. \tag{11}$$

Since $QEH = EH$, (9) and (11) show that EH is a reducing subspace for A and B.

We now restrict the operators A, B, Q to $(EH)^\perp$ and iterate the procedure. We have

$$A = \Sigma \lambda_k E_k,$$

where $\lambda_1, \lambda_2, \ldots$ are the norms of A and its successive restrictions, and $E_1, E_2, \ldots$ are the projections onto the corresponding eigenspaces. By (10), we have $Q^*QE_k = QQ^*E_k = E_k$ $(k = 1, 2, \ldots)$, and so (4) is proved, i.e.

$$Q^*QT^*T = QQ^*T^*T = T^*T. \tag{12}$$

Similarly, starting from $TT^* = Q^*(TQQ^*T^*)Q$, we obtain,

$$Q^*QTT^* = QQ^*TT^* = TT^*. \tag{13}$$

Since $(T^*Tx + TT^*x, x) = \|Tx\|^2 + \|T^*x\|^2$, $((T^*T + TT^*)H)^\perp$ is the reducing null-space of T. Therefore $(T^*T + TT^*)H$ is dense in H, and (12) and (13) give $Q^*Q = QQ^* = I$, i.e. Q is unitary on H.

Since $Q = FG$ with $F, G \in B_1(H)$, we now have

$$I = Q^*Q = G^*F^*FG \leq G^*G \leq I,$$

and so $G^*G = I$. Similarly from $QQ^* = I$, we conclude that $FF^* = I$.

These results have been obtained from the equality (1) and the hypothesis that T is a compact linear operator with zero reducing null-space. But we also know that

$$S = (GF)^*S\,GF,$$

and that S is a compact linear operator with zero reducing null-space. Therefore $F^*F = I$ and $GG^* = I$. This completes the proof that F and G are unitary, and so S and T are unitarily equivalent.

Remarks. (1) By Theorem 7, it is enough in Theorem 9 to assume the compactness of one of the operators S, T.

(2) In Theorem 9 we have assumed for simplicity that S, T act on the same Hilbert space H. However, this restriction is easily removed. Let S, T be compact linear operators on Hilbert spaces H, K respectively with zero reducing null-spaces. Since $(S^*SH + SS^*H)^{\perp}$ is the reducing null-space of S, we have $(S^*SH + SS^*H)^{-} = H$, and therefore H is separable. Similarly K is separable, and so there exists a unitary mapping of H onto K and T can be replaced by a unitarily equivalent operator on H.

In the rest of this section we give a brief and very incomplete account of the work of Arveson [102, 103, 104] on matrix ranges. The account is intended to be self-contained but it leaves on one side many of the deeper questions.

Let A be a unital B^*-algebra, and denote by $CP(A, B(\underset{\sim}{C}^n), I_n)$ the set of all completely positive linear mappings ϕ of A into $B(\underset{\sim}{C}^n)$ such that $\phi(1) = I_n$. Complete positivity of ϕ is defined in terms of the positivity of the natural extensions of ϕ to certain tensor products of algebras. We shirk this by assuming as known the following theorem of Stinespring. For the purposes of the present account this theorem may be taken as the definition of $CP(A, B(\underset{\sim}{C}^n), I_n)$.

We recall that a representation of a B^*-algebra A on a Hilbert space K is a star homomorphism of A into $B(K)$.

Theorem 10. (Stinespring [212].) $CP(A, B(\underset{\sim}{C}^n), I_n)$ is the set of all mappings $\phi : A \to B(\underset{\sim}{C}^n)$ of the form

$$\phi(a) = V^* \pi(a) V \qquad (a \in A),$$

where π is a representation of A on a Hilbert space K, V is a linear isometry of $\underset{\sim}{C}^n$ into K, and K is the closed linear span of $\{\pi(a)V\underset{\sim}{\lambda} : \underset{\sim}{\lambda} \in \underset{\sim}{C}^n,\ a \in A\}$.

Remarks. (1) We have $\pi(1) = I$ (the identity operator on K) since

$$\pi(1)\, \pi(a)\, V\underset{\sim}{\lambda} = \pi(1a)\, V\underset{\sim}{\lambda} = \pi(a)\, V\underset{\sim}{\lambda} \qquad (a \in A,\ \underset{\sim}{\lambda} \in \underset{\sim}{C}^n).$$

(2) When $n = 1$, $CP(A, B(\underset{\sim}{C}^n), I_n)$ coincides with the set $D(1)$ of all normalized states on A. To see this, let $n = 1$, let $\phi \in CP(A, \underset{\sim}{C}, I_1)$, and take $u = V1$. Then $\|u\| = 1$, $V\lambda = \lambda u$, V^* is the mapping of K into $\underset{\sim}{C}$ given by $V^*x = (x,u)$ $(x \in K)$; for we have

$$(V^*x,\ \lambda) = (x,\ V\lambda) = (x,\ \lambda u) = (x, u)\lambda^* = ((x, u),\ \lambda).$$

Then $\phi(a)\lambda = V^*\pi(a)V\lambda = (\pi(a)V\lambda, u) = (\pi(a)u, u)\lambda$, and so if we identify $B(\underset{\sim}{C})$ with $\underset{\sim}{C}$ in the usual way, ϕ is the linear functional on A given by $\phi(a) = (\pi(a)u, u)$ $(a \in A)$, i. e. ϕ is a normalized state on A. Conversely, given a normalized state ϕ on A, there exists a Hilbert space K, a representation π of A on K, and a unit vector $u \in K$ such that $\phi(a) = (\pi(a)u, u)$ $(a \in A)$, and u is a topologically cyclic vector.

Definition 11. The nth algebraic matrix range of $T \in B(H)$ is the subset $\mathscr{W}_n(T)$ of $B(\underset{\sim}{C}^n)$ defined by

$$\mathscr{W}_n(T) = \{\phi(T) : \phi \in CP(C^*(T),\ B(\underset{\sim}{C}^n),\ I_n)\},$$

where $C^*(T)$ is the least closed *-subalgebra of $B(H)$ containing T and I.

Our next aim is to express $\mathscr{W}_n(T)$ in the form $W_n(\underset{\sim}{\gamma}(T))$ for a suitably constructed image $\underset{\sim}{\gamma}(T)$ of the operator T.

Notation. Let A be (as before) a unital B*-algebra with unit element 1. Then the set of normalized states on A coincides with the set $D(1) = \{f \in S(A') : f(1) = 1\}$. Given $f \in D(1)$, there is a standard construction of a Hilbert space H_f, a representation π_f of A on H_f, and a topologically cyclic unit vector $u \in H_f$ such that

$$f(a) = (\pi_f(a)u, u) \qquad (a \in A).$$

Conversely, given a topologically cyclic representation of A on a Hilbert space H, we may choose a topologically cyclic unit vector $v \in H$. Then the functional f defined on A by taking

$$f(a) = (\pi(a)v, v) \qquad (a \in A),$$

belongs to D(1), and the representation π on H is unitarily equivalent to the representation π_f on H_f, i.e. there exists a unitary mapping U of H onto H_f such that $\pi_f(a)U = U\pi(a)$ $(a \in A)$.

Given an indexed set $\{H_\lambda : \lambda \in \Lambda\}$ of Hilbert spaces, the <u>Hilbert direct sum</u> $\bigoplus_{\lambda\in\Lambda} H_\lambda$ is the Hilbert space H defined as follows. The elements of H are the functions x on Λ such that $x(\lambda) \in H_\lambda$ and $\sum_{\lambda\in\Lambda} \|x(\lambda)\|^2 < \infty$; the scalar product is defined by

$$(x, y) = \sum_{\lambda\in\Lambda}(x(\lambda), y(\lambda)) \qquad (x, y \in H),$$

the series being absolutely convergent. Given representations $\{\pi_\lambda : \lambda \in \Lambda\}$ of A on the spaces H_λ, the direct sum $\bigoplus_{\lambda\in\Lambda} \pi_\lambda$ is the representation π on $H = \bigoplus_{\lambda\in\Lambda} H_\lambda$ defined by

$$(\pi x)(\lambda) = \pi_\lambda x(\lambda) \qquad (\lambda \in \Lambda, \ x \in H).$$

Since A is a unital B*-algebra every representation τ of A satisfies

$$\|\tau(a)\| \le \|a\| \qquad (a \in A).$$

Thus πx is a well defined element of H.

The <u>universal space</u> for A is the Hilbert direct sum

$$\underset{\sim}{H} = \bigoplus_{f\in D(1)} H_f,$$

and the <u>universal representation</u> is the corresponding representation

$\underset{\sim}{\pi} = \bigoplus_{f \in D(1)} \pi_f.$

The universal space $\underset{\sim}{H}$ is not quite large enough for our purpose, and we therefore construct a space $\underset{\sim}{G}$ which is unitarily equivalent to a Hilbert direct sum of countably many copies of $\underset{\sim}{H}$.

Definition 12. Let $\Lambda = \underset{\sim}{N} \times D(1)$, where $\underset{\sim}{N}$ is the set of all positive integers and $D(1)$ is the set of normalized states on A. For each $\lambda = (n, f) \in \Lambda$, let $G_\lambda = H_f$ and $\gamma_\lambda = \pi_f$. The great universal space $\underset{\sim}{G}$ for A is the Hilbert direct sum

$$\underset{\sim}{G} = \bigoplus_{\lambda \in \Lambda} G_\lambda ,$$

and the great universal representation of A is the representation

$$\underset{\sim}{\gamma} = \bigoplus_{\lambda \in \Lambda} \gamma_\lambda$$

of A on $\underset{\sim}{G}$.

Lemma 13. *Let $\underset{\sim}{M}$ be a subset of $\underset{\sim}{N}$, for each $r \in \underset{\sim}{M}$ let π_r be a topologically cyclic representation of A on a Hilbert space H_r, and let $H = \bigoplus_{r \in \underset{\sim}{M}} H_r$, $\pi = \bigoplus_{r \in \underset{\sim}{M}} \pi_r$.*

Then there exists a closed linear subspace E of the great universal space $\underset{\sim}{G}$ such that

(i) *E is invariant for the great universal representation $\underset{\sim}{\gamma}$,*

(ii) *π is unitarily equivalent to $\underset{\sim}{\gamma}|_E$,*

(iii) $\underset{\sim}{x}(n, f) = 0 \quad (\underset{\sim}{x} \in E,\ n \in \underset{\sim}{N} \backslash \underset{\sim}{M},\ f \in D(1)).$

Proof. Since π_r is a topologically cyclic representation of A, there exists $f_r \in D(1)$ and a unitary mapping U_r of H_r onto H_{f_r} such that

$$U_r(\pi_r(a)x) = \pi_{f_r}(a)(U_r x) \quad (a \in A,\ x \in H_r).$$

Let $\Gamma = \{(r, f_r) : r \in \underset{\sim}{M}\}$, and take

$$E = \{\underset{\sim}{x} \in \underset{\sim}{G} : \underset{\sim}{x}(\lambda) = 0 \ (\lambda \in \Lambda \backslash \Gamma)\}.$$

Since $(\underset{\sim}{\gamma}(a)\underset{\sim}{x})(\lambda) = \gamma_\lambda(a)(\underset{\sim}{x}(\lambda))$, it is clear that E is a closed linear subspace of $\underset{\sim}{G}$ invariant for $\underset{\sim}{\gamma}$. It is also clear that (iii) holds.

Each element of H is a function h on $\underset{\sim}{M}$ with $\|h\|^2 = \sum_{r \in \underset{\sim}{M}} \|h(r)\|^2 < \infty$, and with $h(r) \in H_r$ $(r \in \underset{\sim}{M})$. We define U on H by taking

$$(Uh)(\lambda) = \begin{cases} 0 & (\lambda \in \Lambda\backslash\Gamma), \\ U_r(h(r)) & (\lambda = (r, f_r) \in \Gamma). \end{cases}$$

Plainly

$$\|Uh\|^2 = \sum_{r \in \underset{\sim}{M}} \|U_r(h(r))\|^2 = \sum_{r \in \underset{\sim}{M}} \|h(r)\|^2 = \|h\|^2,$$

and so U is a unitary mapping of H onto E. Let $h \in H$ and $a \in A$. Then, for $\lambda = (r, f_r) \in \Gamma$,

$$\begin{aligned}(U\pi(a)h)(\lambda) &= U_r((\pi(a)h)(r)) \\ &= U_r(\pi_r(a)h(r)) \\ &= \pi_{f_r}(a)(U_r h(r)) \\ &= \pi_{f_r}(a)((Uh)(r, f_r)) = (\underset{\sim}{\gamma}(a)Uh)(\lambda),\end{aligned}$$

and when $\lambda \in \Lambda\backslash\Gamma$,

$$(U\pi(a)h)(\lambda) = 0 = (\underset{\sim}{\gamma}(a)Uh)(\lambda).$$

Thus

$$U\pi(a)h = \underset{\sim}{\gamma}(a)Uh \qquad (a \in A,\ h \in H),$$

and so π is unitarily equivalent to $\underset{\sim}{\gamma}|_E$.

Lemma 14. Let $\phi \in CP(A, B(\underset{\sim}{C}^n), I_n)$. Then there exist $m \in \underset{\sim}{N}$ with $m \le n$, Hilbert spaces $K_1, \ldots, K_m$, topologically cyclic representations τ_j of A on K_j $(j = 1, \ldots, m)$, and a linear isometry Q of $\underset{\sim}{C}^n$ into $\bigoplus_{j=1}^{m} K_j$ such that

$$\phi(a) = Q^* \bigoplus_{j=1}^{m} \tau_j(a)Q \qquad (a \in A).$$

Proof. By Theorem 10, there exist a Hilbert space K, a representation π of A on K and a linear isometry V of $\underset{\sim}{C}^n$ into K such that

$$\phi(a) = V^*\pi(a)V \quad (a \in A),$$

and K is the closed linear span of $\pi(A)L$, where L is the n-dimensional subspace $V\underset{\sim}{C}^n$ of K.

Choose $u_1 \in L\setminus\{0\}$, and take $K_1 = (\pi(A)u_1)^-$. If $L \subset K_1$, then $K_1 = K$ and the lemma is proved with $m = 1$. If $L \not\subset K_1$, let P_2 be the projection onto $K_1^\perp$. Then $P_2L \neq \{0\}$. Choose $u_2 \in L$ such that $P_2u_2 = v_2 \neq 0$, and take $K_2 = \{\pi(A)v_2\}^-$. Since $K_1^\perp$ is invariant for π and $v_2 \in K_1^\perp$, we have $K_2 \subset K_1^\perp$. If $L \subset K_1 \oplus K_2$, then $K = K_1 \oplus K_2$ and the theorem is proved with $m = 2$. If not, we let P_3 be the projection onto $(K_1 \oplus K_2)^\perp$, and choose $u_3 \in L$ such that $P_3u_3 = v_3 \neq 0$, and so on.

The vectors $u_1, u_2, \ldots$ in L obtained in this way are linearly independent; for $P_k \geq P_j$ $(k < j)$, and so if

$$\lambda_1u_1 + \lambda_2u_2 + \ldots + \lambda_ju_j = 0,$$

we have $P_ju_k = P_jP_ku_k = P_jv_k = 0$ $(k < j)$, from which $\lambda_jv_j = \lambda_jP_ju_j = 0$. Therefore there exists $m \leq n$ such that $L \subset K_1 \oplus \ldots \oplus K_m$. Then $K = K_1 \oplus \ldots \oplus K_m$, and the lemma is proved (using the obvious unitary equivalence of an orthogonal sum of closed subspaces to the corresponding Hilbert direct sum).

Theorem 15. Let $\phi \in CP(A, B(\underset{\sim}{C}^n), I_n)$ and let $\underset{\sim}{G}$, $\underset{\sim}{\gamma}$ denote the great universal space and representation for A. Then there exists $m \in \underset{\sim}{N}$ with $m \leq n$ such that for every subset $\underset{\sim}{M}$ of $\underset{\sim}{N}$ with m elements, there exists a closed linear subspace E of $\underset{\sim}{G}$ and a linear isometry V of $\underset{\sim}{C}^n$ into E satisfying

(i) E is invariant for $\underset{\sim}{\gamma}$,

(ii) $\phi(a) = V^*\underset{\sim}{\gamma}(a)V \quad (a \in A)$,

(iii) $\underset{\sim}{x}(n, f) = 0 \quad (\underset{\sim}{x} \in E,\ n \in \underset{\sim}{N}\backslash\underset{\sim}{M},\ f \in D(1))$.

Conversely, each mapping $\phi : A \to B(\underset{\sim}{C}^n)$ of the form (ii) with V an isometry of $\underset{\sim}{C}^n$ into $\underset{\sim}{G}$ belongs to $CP(A, B(\underset{\sim}{C}^n), I_n)$.

Proof. Let $m, K_1, \ldots, K_m, \tau_1, \ldots, \tau_m, Q$ be as in Lemma 14, and let $\underset{\sim}{M}$ be any subset of $\underset{\sim}{N}$ with m elements. Let $\underset{\sim}{M} = \{k_1, k_2, \ldots, k_m\}$, take $H_{k_i} = K_i,\ \pi_{k_i} = \tau_i \quad (i = 1, \ldots, m)$ and let

$$H = \bigoplus_{r \in \underset{\sim}{M}} H_r, \qquad \pi = \bigoplus_{r \in \underset{\sim}{M}} \pi_r .$$

Then Q is an isometry of $\underset{\sim}{C}^n$ into H, and $\phi(a) = Q^*\pi(a)Q \quad (a \in A)$. By Lemma 13, there exists a subspace E of $\underset{\sim}{G}$ such that (i) and (iii) hold, and π is unitarily equivalent to $\underset{\sim}{\gamma}|_E$. Thus there exists a unitary mapping U of H into E such that

$$U\pi(a)h = \underset{\sim}{\gamma}(a)Uh \qquad (a \in A,\ h \in H).$$

Let $V = UQ$. Then V is an isometry of $\underset{\sim}{C}^n$ into E, and

$$\begin{aligned} V^*\underset{\sim}{\gamma}(a)V &= Q^*U^*\underset{\sim}{\gamma}(a)UQ \\ &= Q^*U^*U\pi(a)Q \\ &= Q^*\pi(a)Q \\ &= \phi(a) \qquad (a \in A). \end{aligned}$$

The converse is obvious (Theorem 10).

Definition 16. $\mathscr{W}_n(A, a) = \{\phi(a) : \phi \in CP(A, B(\underset{\sim}{C}^n), I_n)\}$.

Corollary 17. $\mathscr{W}_n(A, a) = W_n(\underset{\sim}{\gamma}(a)) \quad (a \in A,\ n \in \underset{\sim}{N})$, where $\underset{\sim}{\gamma}$ is the great universal representation of A.

Corollary 18. $\mathscr{W}_n(T) = W_n(\underset{\sim}{\gamma}(T)) \quad (T \in B(H),\ n \in \underset{\sim}{N})$, where $\underset{\sim}{\gamma}$ is the great universal representation of $C^*(T)$.

Definition 19. A subset K of $B(\underset{\sim}{C}^n)$ is matrix convex if $r \in \underset{\sim}{N},\ S_1, \ldots, S_r \in K,\ A_1, \ldots, A_r \in B(\underset{\sim}{C}^n)$,

$A_1^*A_1 + \ldots + A_r^*A_r = I_n \Rightarrow A_1^*S_1A_1 + \ldots + A_r^*S_rA_r \in K.$

Corollary 20. $\mathscr{W}_n(A, a)$ <u>is matrix convex.</u>

Proof. Let $S_1, \ldots, S_r \in \mathscr{W}_n(A, a)$, $A_1, \ldots, A_r \in B(\underset{\sim}{C}^n)$, $A_1^*A_1 + \ldots + A_r^*A_r = I_n$. There exist $\phi_j \in CP(A, B(\underset{\sim}{C}^n), I_n)$ such that $S_j = \phi_j(a)$ $(j = 1, \ldots, r)$. Then there exist closed linear subspaces E_j of $\underset{\sim}{G}$, mutually orthogonal and invariant for $\underset{\sim}{\gamma}$, and linear isometries V_j of $\underset{\sim}{C}^n$ into E_j such that $\phi_j(a) = V_j^*\underset{\sim}{\gamma}(a)V_j$ $(j = 1, \ldots, r)$. Let $V = V_1A_1 + \ldots + V_rA_r$. Since the subspaces E_j are mutually orthogonal and invariant for $\underset{\sim}{\gamma}$, we have

$$V_i^*V_j = 0, \quad V_i^*\underset{\sim}{\gamma}(a)V_j = 0 \quad (i \neq j), \quad V_i^*V_i = I_n.$$

Therefore

$$V^*V = A_1^*V_1^*V_1A_1 + \ldots + A_r^*V_r^*V_rA_r = I_n,$$

$$V^*\underset{\sim}{\gamma}(a)V = A_1^*V_1^*\underset{\sim}{\gamma}(a)V_1A_1 + \ldots + A_r^*V_r^*\underset{\sim}{\gamma}(a)V_rA_r = A_1^*S_1A_1 + \ldots + A_r^*S_rA_r.$$

Thus $A_1^*S_1A_1 + \ldots + A_r^*S_rA_r \in \mathscr{W}_n(A, a)$.

We aim to show that $\mathscr{W}_n(A, a)$ is independent of the choice of B*-algebra containing a and 1. This will follow at once from the following theorem.

Theorem 21. <u>Let</u> B <u>be a unital</u> B*-<u>algebra and let</u> A <u>be a closed</u> *-<u>subalgebra of</u> B <u>containing</u> 1. <u>Then the restriction mapping</u> $\phi \to \phi|_A$ <u>maps</u> $CP(B, B(\underset{\sim}{C}^n), I_n)$ <u>onto</u> $CP(A, B(\underset{\sim}{C}^n), I_n)$.

That the restriction mapping maps $CP(B, B(\underset{\sim}{C}^n), I_n)$ into $CP(A, B(\underset{\sim}{C}^n), I_n)$ is obvious, by Theorem 10. For given a representation π of B, $\pi|_A$ is a representation of A. To prove that the mapping is surjective, we use the following lemma.

Lemma 22. <u>With</u> A, B <u>as in Theorem 21, let</u> τ <u>be a topologically cyclic representation of</u> A <u>on a Hilbert space</u> K. <u>Then there exists a topologically cyclic representation</u> π <u>of</u> B <u>on a Hilbert space</u> H <u>and a closed linear subspace</u> H_0 <u>of</u> H <u>invariant for</u> $\pi|_A$ <u>such that</u> $\pi|_A$ <u>is</u>

a topologically cyclic representation of A on H_0 unitarily equivalent to τ.

Proof. There exists a topologically cyclic unit vector $x_0 \in K$. Define

$$f(a) = (\tau(a)x_0, x_0) \qquad (a \in A).$$

Then $f \in D(A, 1)$. By the Hahn-Banach theorem there exists $g \in D(B, 1)$ with $g|_A = f$. Let π, H denote the representation of B and Hilbert space given by the state g (i. e. $\pi = \pi_g$, $H = H_g$). There exists a topologically cyclic unit vector $y_0 \in H$ with

$$g(b) = (\pi(b)y_0, y_0) \qquad (b \in B).$$

Let $H_0 = (\pi(A)y_0)^-$. Then H_0 is invariant for $\pi|_A$, and $\pi|_A$ has the topologically cyclic vector $y_0 \in H_0$. We have

$$(\tau(a)x_0, x_0) = f(a) = g(a) = (\pi(a)y_0, y_0) \qquad (a \in A),$$

and therefore

$$(\tau(a)x_0, \tau(a')x_0) = (\pi(a)y_0, \pi(a')y_0) \qquad (a, a' \in A).$$

Define $U_0 : \tau(A)x_0 \rightarrow \pi(A)y_0$, by

$$U_0(\tau(a)x_0) = \pi(a)y_0.$$

This is a well-defined isometric linear mapping of $\tau(A)x_0$ onto $\pi(A)y_0$, and hence extends to a unitary mapping U of K onto H_0. Moreover, since

$$(\tau(a)\tau(a_1)x_0, \tau(a_2)x_0) = (\pi(a)\pi(a_1)y_0, \pi(a_2)y_0),$$

we have

$$(\tau(a)x, x') = (\pi(a)Ux, Ux'),$$

and so

$$\tau(a) = U^*\pi(a)U \qquad (a \in A).$$

Proof of Theorem 21. Let $\phi \in CP(A, B(\underset{\sim}{C}^n), I_n)$, let m, $K_1, \ldots, K_m, \tau_1, \ldots, \tau_m, Q$ be as in Lemma 14. By Lemma 22 there exist topologically cyclic representations π_j of B on Hilbert spaces H_j and closed linear subspaces H_{j0} invariant for $\pi_j|_A$ such that $\pi_j|_A$ is a topologically cyclic representation of A on H_{j0} unitarily equivalent to τ_j, i. e. there exists a unitary mapping U_j of K_j onto H_{j0} with

$$\tau_j(a) = U_j^* \pi_j(a)U_j \qquad (a \in A). \tag{14}$$

Let P_j be the projection of H_j onto H_{j0}, let $\tau = \bigoplus_{j=1}^{m} \tau_j$, $K = \bigoplus_{j=1}^{m} K_j$,

$$H = \bigoplus_{j=1}^{m} H_j, \quad H_0 = \bigoplus_{j=1}^{m} H_{j0}, \quad \pi = \bigoplus_{j=1}^{m} \pi_j, \quad P = \bigoplus_{j=1}^{m} P_j, \quad U = \bigoplus_{j=1}^{m} U_j.$$

Then π is a representation of B on H, H_0 is an invariant subspace for $\pi|_A$, P is the projection of H onto H_0, and U is a unitary mapping of K onto H_0. We define $\psi : B \rightarrow B(\underset{\sim}{C}^n)$ by taking

$$\psi(b) = Q^*U^*P\,\pi(b)\,PUQ \qquad (b \in B).$$

Since P is the identity on $UQ\underset{\sim}{C}^n$, PUQ is an isometry of $\underset{\sim}{C}^n$ into H, and so $\psi \in CP(B, B(\underset{\sim}{C}^n), I_n)$. Also, by Lemma 14,

$$\phi(a) = Q^*\,\tau(a)Q\,,$$

and by (14)

$$\tau(a) = \bigoplus_{j=1}^{n} U_j^*\,\pi_j(a)\,U_j = U^*\,\pi(a)\,U\,.$$

Thus, for $a \in A$,

$$\begin{aligned}\phi(a) &= Q^*U^*\,\pi(a)\,UQ\\ &= Q^*U^*P\,\pi(a)\,PUQ\\ &= \psi(a).\end{aligned}$$

We have proved that $\phi = \psi|_A$, and so the theorem is proved.

Corollary 23. *Let* A, B *be as in Theorem* 21. *Then*

$$\mathscr{W}_n(A, a) = \mathscr{W}_n(B, a) \qquad (a \in A).$$

Corollary 24. *Let* A_1, A_2 *be closed* *-*subalgebras of* A *containing* 1. *Then*

$$\mathscr{W}_n(A_1, a) = \mathscr{W}_n(A_2, a) \qquad (a \in A_1 \cap A_2) .$$

Proof. $A_1 \cap A_2$ is a closed *-subalgebra of A containing 1. Apply Corollary 23.

Remarks. (1) In [103], Arveson proves that $\mathscr{W}_n(T)$ is compact. Can this be deduced from Theorem 15?

(2) Can a proof of Arveson's theorem [104], that the $\mathscr{W}_n(T)$ form a complete set of unitary invariants for irreducible compact operators, be based on the elementary methods developed above?

37. SOME AXIOMS FOR NUMERICAL RANGES OF OPERATORS

In these lecture notes we have been concerned with several numerical ranges of an operator T on a Banach space: the algebra numerical range V(B(X), T), the spatial numerical range V(T), the numerical ranges W(T) corresponding to the semi-inner-products on X that determine the norm of X, and, when X is a dual space of a Banach space, the lower numerical range LV(T) (Definition 17.1). For some purposes these ranges are interchangeable, and it therefore seems appropriate to round off this volume by introducing an abstract concept that includes these and some other ranges that have been found useful in the study of matrices.

Notation. X will denote a normed linear space over $\underset{\sim}{C}$. Given $T \in B(X)$, we denote by $|T|$ its norm, by $\rho(T)$ its spectral radius, and by apSp(T) its approximate point spectrum, i.e. the set of all $\lambda \in \underset{\sim}{C}$ such that

$$\inf \{ \|(\lambda I - T)x\| : x \in S(X) \} = 0.$$

L will denote a complex linear subspace of B(X) containing the identity operator I.

We denote by $\mathscr{C}$ the set of non-void subsets of $\underset{\sim}{C}$, and given $E \in \mathscr{C}$, we write $|E|$, Re E to denote the sets $\{|\lambda| : \lambda \in E\}$, $\{\mathrm{Re}\,\lambda : \lambda \in E\}$ respectively.

Definition 1. A printer on L is a mapping $\Phi : L \to \mathscr{C}$ such that the following axioms hold:

P(i). $\Phi(\alpha I + \beta T) = \alpha + \beta\Phi(T) \qquad (\alpha, \beta \in \underset{\sim}{C},\ T \in L);$

P(ii). $\sup|\Phi(T)| \le |T| \qquad (T \in L);$

P(iii). $\inf|\Phi(T)| \le \|Tx\| \qquad (T \in L,\ x \in S(X)).$

The set $\Phi(T)$ is called the print of T.

Remarks. (1) The mappings $T \to V(B(X), T)$, $T \to V(T)$, $T \to W(T)$ are easily seen to be printers on B(X). To see that $T \to LV(T)$ is also a printer, let X be the dual space of a Banach space Y, so that $LV(T) = \{(Tf)(y) : (y, f) \in \Pi(Y)\}$. Given $x \in S(X)$, the Bishop-Phelps theorem allows us to approximate x in norm by $f \in S(X)$ such that $(y, f) \in \Pi(Y)$. Then $\|Tf\| \ge |(Tf)(y)|$ and $(Tf)(y) \in LV(T)$. This shows that axiom P(iii) is satisfied, and P(i) and P(ii) are obvious.

(2) The axioms P(ii) and P(iii) can be written in the symmetrical form:

P(ii). $\sup|\Phi(T)| \le \sup\{\|Tx\| : x \in S(X)\},$

P(iii). $\inf|\Phi(T)| \le \inf\{\|Tx\| : x \in S(X)\}.$

(3) Given a printer Φ on a linear subspace L of B(X), Φ can be extended with a great deal of freedom to a printer on B(X). In fact, let Ψ be any printer on B(X) and define Φ_1 on B(X) by

$$\Phi_1(T) = \begin{cases} \Phi(T) & (T \in L) \\ \Psi(T) & (T \in B(X)\backslash L). \end{cases}$$

For the axioms P(ii) and P(iii) involve only a single operator, and for axiom P(i) it is enough to note that if $\alpha I + \beta T \in L$, then either $\beta = 0$ or $T \in L$.

(4) We may take the space X to be complete whenever this is convenient. To see this, let $\hat{X}$ denote the completion of the normed linear space X. Each $T \in B(X)$ has a unique extension $\hat{T} \in B(\hat{X})$. Given a printer Φ on L, let $\hat{\Phi}$ be defined on $\hat{L}$ by

$$\hat{\Phi}(\hat{T}) = \Phi(T) \qquad (T \in L).$$

Then $\hat{\Phi}$ is a printer on $\hat{L}$. Axioms P(i), (ii) are obviously satisfied since $T \rightarrow \hat{T}$ is a linear isometry. To prove P(iii), let $x \in S(\hat{X})$. Then there exist $x_n \in S(X)$ with $x = \lim_{n\to\infty} x_n$, and we have $\hat{T}x = \lim_{n\to\infty} Tx_n$. Then, since

$$\inf|\Phi(T)| \le \|Tx_n\| \qquad (n = 1, 2, \ldots),$$

we have

$$\inf|\hat{\Phi}(\hat{T})| \le \|\hat{T}x\|.$$

Lemma 2. Let Φ be a printer on L, let $T \in L$, and let $\lambda \in \underset{\sim}{C}$.

(i) If $d(\lambda, \Phi(T)) = \delta$, then

$$\|(\lambda I - T)x\| \ge \delta\|x\| \qquad (x \in X).$$

(ii) $\operatorname{apSp}(T) \subset (\Phi(T))^-$.

(iii) $\rho(T) \le \sup|\Phi(T)|$.

Proof. (i) Let $d(\lambda, \Phi(T)) = \delta$. We have $\lambda I - T \in L$, and so by axiom P(iii), for all $x \in S(X)$,

$$\|(\lambda I - T)x\| \ge \inf|\Phi(\lambda I - T)| = \inf|\lambda - \Phi(T)| = \delta.$$

(ii) Let $\lambda \in \operatorname{apSp}(T)$. Then $\inf\{\|(\lambda I - T)x\| : x \in S(X)\} = 0$, and so, in (i), $\delta = 0$, i.e. $\lambda \in (\Phi(T))^-$.

(iii) By Remark (4) above, we may assume that X is complete. Then $\partial\operatorname{Sp}(T) \subset \operatorname{apSp}(T)$, and so, by (ii), $\rho(T) \le \sup|\Phi(T)|$.

Lemma 3. Let Φ be a printer on L and let $T \in L$. Then

$$\sup \operatorname{Re} \Phi(T) = \inf_{\alpha>0} \frac{1}{\alpha}(|I+\alpha T|-1) = \lim_{\alpha\to 0+} \frac{1}{\alpha}(|I+\alpha T| - 1).$$

Proof. This is a slight recasting of a familiar argument (NRI Theorem 2.5 and NRI Lemma 9.2). Let $\mu = \sup \operatorname{Re} \Phi(T)$. Given $\lambda \in \Phi(T)$ and $\alpha > 0$, we have $\lambda \in \frac{1}{\alpha}\{\Phi(I + \alpha T) - 1\}$, and so

$$\operatorname{Re}\lambda \le \frac{1}{\alpha}\{\sup \operatorname{Re} \Phi(I+\alpha T)-1\} \le \frac{1}{\alpha}\{\sup|\Phi(I+\alpha T)|-1\} \le \frac{1}{\alpha}\{|I+\alpha T|-1\}.$$

Thus

$$\mu \le \inf_{\alpha>0} \frac{1}{\alpha}(|I + \alpha T| - 1). \tag{1}$$

Assume now that $T \neq 0$, the case $T = 0$ being trivial, and let $0 < \alpha < |T|^{-1}$. Since $\mu \le |T|$, we have $d(\alpha^{-1}, \Phi(T)) \ge \alpha^{-1} - \mu$. Therefore, by Lemma 2,

$$\|(\alpha^{-1}I - T)x\| \ge (\alpha^{-1} - \mu)\ \|x\| \qquad (x \in X),$$

i.e.

$$\|(I - \alpha T)x\| \ge (1 - \alpha\mu)\ \|x\| \qquad (x \in X).$$

Replacing x by $(I + \alpha T)x$, we obtain

$$\|(I + \alpha T)x\| \le (1 - \alpha\mu)^{-1}\ \|(I - \alpha^2 T^2)x\| \qquad (x \in X),$$

and so

$$|I + \alpha T| \le (1 - \alpha\mu)^{-1}\ (1 + \alpha^2|T|^2),$$

$$\frac{1}{\alpha}\{|I + \alpha T| - 1\} \le (1 - \alpha\mu)^{-1}\ (\mu + \alpha|T|^2).$$

With (1), this completes the proof.

Theorem 4. Let Φ be a printer on L, and let $T \in L$. Then

$\overline{\mathrm{co}}\ \Phi(T) = V(B(X), T)$.

Proof. For all $\alpha, \beta \in \mathbb{C}$, we have $\alpha I + \beta T \in L$ and

$$\Phi(\alpha I + \beta T) = \alpha + \beta\Phi(T).$$

Lemma 3 now shows that all prints of T have the same closed convex hull, namely $V(B(X), T)$.

Remarks. (1) This theorem shows that a large part of numerical range theory is valid for a general printer, in particular all results concerning the numerical radius. Clearly also every printer on $B(X)$ determines the same class of Hermitian operators (those with their prints in $\underset{\sim}{R}$).

(2) Certain elementary properties of the usual numerical ranges do not appear. For example, we have no reason to suppose that

$$\Phi(A + B) \subset \Phi(A) + \Phi(B).$$

This does hold whenever there exists a subset Γ of $D(I)$ such that

$$\Phi(T) = \{f(T) : f \in \Gamma\} \qquad (T \in L).$$

We therefore have the problem of characterizing abstractly those printers that are of this form for some subset Γ of $D(I)$. In the special case when $L = [I, A]$, the linear span of I and A, every printer on L is of this form. For let $L = [I, A]$, let Φ be a printer on L and let $\lambda \in \Phi(A)$. Then $\alpha + \beta\lambda \in \Phi(\alpha I + \beta A)$, and so

$$|\alpha + \beta\lambda| \leq |\alpha I + \beta A| \qquad (\alpha, \beta \in \underset{\sim}{C}).$$

We may therefore define a linear functional f_0 on L by taking $f_0(\alpha I + \beta A) = \alpha + \beta\lambda$, and we have $\|f_0\| = 1 = f_0(I)$. By the Hahn-Banach theorem, we can extend f_0 to obtain $f \in D(I)$, and have $f(\alpha I+\beta A)=\alpha+\beta\lambda$ $(\alpha, \beta \in \underset{\sim}{C})$.

The following construction yields a class of printers. Let X be a linear space over $\underset{\sim}{C}$, and let Σ be a subset of the algebraic dual X^* of X that separates the points of X and is weak* compact. Define p_Σ, S_Σ, Γ_Σ in turn by

$$p_\Sigma(x) = \sup\{|f(x)| : f \in \Sigma\} \qquad (x \in X),$$

$$S_\Sigma = \{x \in X : p_\Sigma(x) = 1\},$$

$$\Gamma_\Sigma = \{(x, f) \in S_\Sigma \times \Sigma : f(x) = 1\}.$$

Then p_Σ is a norm on X, and we may consider the set $B_\Sigma(X)$ of linear operators on X that are bounded with respect to the norm p_Σ. Given $T \in B_\Sigma(X)$, let $V_\Sigma(T)$ be defined by

$$V_\Sigma(T) = \{f(Tx) : (x, f) \in \Gamma_\Sigma\}.$$

Then V_Σ is a printer on B_Σ (with respect to the norm p_Σ). To prove P(iii), let $x \in S_\Sigma$. Since Σ is weak* compact, the supremum of $\{|f(x)| : f \in \Sigma\}$ is attained. Thus there exists $f \in \Sigma$ with $|f(x)| = p_\Sigma(x) = 1$. Let ζ be the complex conjugate of $f(x)$. Then $(\zeta x, f) \in \Gamma_\Sigma$, $f(T(\zeta x)) \in V_\Sigma(T)$, and so

$$\inf |V_\Sigma(T)| \leq |f(Tx)| \leq p_\Sigma(Tx).$$

In the case when X is a finite dimensional space, numerical ranges closely related to the printers V_Σ have been developed by Bauer [107], Zenger-Deutsch [222], Vogg [217].

Example 5. Given an $n \times n$ matrix (t_{ij}), the <u>Gershgorin discs</u> $G_1, \ldots, G_n$ for the matrix are the discs

$$G_k = t_{kk} + r_k \Delta,$$

where $r_k = \sum_{j=1}^{n} |t_{kj}| - |t_{kk}|$, and $\Delta = \{z \in \underset{\sim}{C} : |z| \leq 1\}$. The union of the Gershgorin discs is a print $V_\Sigma(T)$ of the kind considered above. To see this, take $X = \underset{\sim}{C}^n$, and let $f_1, \ldots, f_n$ denote the coordinate functionals on X, i.e.

$$f_k(\underset{\sim}{x}) = x_k \quad (\underset{\sim}{x} = (x_1, \ldots, x_n) \in \underset{\sim}{C}^n).$$

Take $\Sigma = \{f_1, f_2, \ldots, f_n\}$. Then

$$p_\Sigma(\underset{\sim}{x}) = \max\{|x_1|, \ldots, |x_n|\} = \|\underset{\sim}{x}\|_\infty \quad (\underset{\sim}{x} \in \underset{\sim}{C}^n),$$

and $(\underset{\sim}{x}, f) \in \Gamma_\Sigma$ if and only if $f = f_k$ for some $k \in \{1, \ldots, n\}$, and

$$x_k = 1 = \max\{|x_j| : 1 \le j \le n\}.$$

Let T be the linear operator on $\underset{\sim}{C}^n$ with the matrix representation (t_{ij}) relative to the natural basis. Then for $(\underset{\sim}{x}, f) \in \Gamma_\Sigma$ as above, we have

$$f(T\underset{\sim}{x}) = f_k(T\underset{\sim}{x}) = \sum_{j=1}^{n} t_{kj}x_j$$

$$= t_{kk} + \sum_{j \ne k} t_{kj}x_j.$$

Clearly $|f(T\underset{\sim}{x}) - t_{kk}| \le r_k$, so that $f(T\underset{\sim}{x}) \in G_k$; and also all points of G_k can be obtained by appropriate choice of x_j $(j \ne k)$. Thus

$$V_\Sigma(T) = G_1 \cup G_2 \cup \dots \cup G_n.$$

It follows at once from Lemma 2 (ii) that all eigenvalues of the matrix are contained in the union of the Gershgorin discs.

In seeking information about the eigenvalues, or more generally the approximate point spectrum, of an operator it is clearly desirable to use as small a print as possible. On the other hand some of the very interesting theorems that are valid for the spatial numerical range of an operator fail for some small prints. In order to recover these theorems we may strengthen the axioms of a printer as follows. We return to the situation in which X is a normed linear space over $\underset{\sim}{C}$ and L is a linear subspace of $B(X)$ containing I.

Definition 6. A <u>full printer</u> on L is a mapping $\Phi : L \to \mathscr{C}$ that satisfies the axioms P(i), P(ii), and

F. P. $\inf|\Phi(T)| \le |f(Tx)| \qquad (T \in L,\ (x, f) \in \Pi(X)).$

Theorem 7. <u>Let Φ be a full printer on L and let $T \in L$. Then</u>

$$V(T) \subset (\Phi(T))^-.$$

Proof. If $0 \in V(T)$, there exists $(x, f) \in \Pi(X)$ with $f(Tx) = 0$. Then $\inf|\Phi(T)| \le 0$, and so $0 \in (\Phi(T))^-$. Given $\lambda \in V(T)$, we have

$0 \in V(\lambda I - T)$, and therefore $0 \in (\Phi(\lambda I - T))^- = (\lambda - \Phi(T))^-$, $\lambda \in (\Phi(T))^-$.

Corollary 8. Let X be a Banach space, let Φ be a full printer on L, and let $T \in L$. Then

$$\mathrm{coSp}(T) \subset (\Phi(T))^-.$$

Proof. Theorem 7 and the theorem of Zenger-Crabb (Theorem 19.4).

We have resisted the temptation to introduce axioms for essential numerical ranges, matrix ranges, and numerical ranges of Banach algebra elements other than $V(A, a)$. Nor have we considered the numerical ranges of non-linear mappings. In this connection we should draw attention to Harris [159] in which a successful theory of numerical ranges of analytic mappings is developed.

Bibliography

(The first seven items complete the details for some of the references in NRI.)

17. M. J. Crabb, 'Numerical range estimates for the norms of iterated operators', Glasgow Math. J. 11 (1970), 85-7.

21. J. Duncan, C. M. McGregor, J. D. Pryce, and A. J. White, 'The numerical index of a normed space', J. London Math. Soc. (2) 2 (1970), 481-8.

28. B. W. Glickfeld, 'On an inequality of Banach algebra geometry and semi-inner-product space theory', Illinois J. Math. 14 (1970), 76-81.

53. A. L. T. Paterson, 'Isometries between B*-algebras', Proc. Amer. Math. Soc. 22 (1970), 570-2.

61. S. Shirali and J. W. M. Ford, 'Symmetry in complex involutory Banach algebras, II', Duke Math. J. 37 (1970), 275-80.

62. A. M. Sinclair, 'The norm of a Hermitian element in a Banach algebra', Proc. Amer. Math. Soc. 28 (1971), 446-50.

63. A. M. Sinclair, 'Jordan homomorphisms and derivations on semi-simple Banach algebras', Proc. Amer. Math. Soc. 24 (1970), 209-14.

100. J. H. Anderson, 'Derivations, commutators and the essential numerical range', Ph. D. thesis, Indiana University, 1971.

101. J. H. Anderson and J. G. Stampfli, 'Commutators and compressions', Israel Math. J. 10 (1971), 433-41.

102. W. B. Arveson, 'Subalgebras of C*-algebras', Acta Math. 123 (1969), 141-224.

103. W. B. Arveson, 'Subalgebras of C*-algebras II', Acta Math. 128 (1972), 271-308.

104. W. B. Arveson, 'Unitary invariants for compact operators', Bull. Amer. Math. Soc. 76 (1970), 88-91.

105. E. Asplund and V. Pták, 'A minimax inequality for operators and a related numerical range', Acta Math. 126 (1971), 53-62.

106. G. de Barra, J. R. Giles, and B. Sims, 'On the numerical range of compact operators on Hilbert space', J. London Math. Soc. (2) 5 (1972), 704-706.

107. F. L. Bauer, 'Fields of values and Gershgorin discs', Numer. Math. 12 (1968), 91-5.

108. S. K. Berberian, 'The numerical range of a normal operator', Duke Math. J. 31 (1964), 479-83.

109. E. Berkson, 'Hermitian projections and orthogonality in Banach spaces', Proc. London Math. Soc. (3) 24 (1972), 101-18.

110. E. Berkson, 'Action of W*-algebras on Banach spaces', Math. Ann. 189 (1970), 261-71.

111. E. Berkson, 'A characterization of complex Hilbert spaces', Bull. London Math. Soc. 2 (1970), 313-5.

112. E. Berkson and H. R. Dowson, 'Prespectral operators', Illinois J. Math. 13 (1969), 291-315.

113. E. Berkson, H. R. Dowson, and G. A. Elliott, 'On Fuglede's theorem and scalar-type operators', Bull. London Math. Soc. 4 (1972), 13-16.

114. A. S. Besicovitch, Almost periodic functions, Dover publications, 1954.

115. B. Bollobás, 'An extension to the theorem of Bishop and Phelps', Bull. London Math. Soc. 2 (1970), 181-2.

116. B. Bollobás, 'The power inequality on Banach spaces', Proc. Camb. Phil. Soc. 69 (1971), 411-15.

117. B. Bollobás, 'The numerical range in Banach algebras and complex functions of exponential type', Bull. London Math. Soc. 3 (1971), 27-33.

118. B. Bollobás, 'A property of Hermitian elements', J. London Math. Soc. (2) 4 (1971), 379-80.

119. F. F. Bonsall, 'Compact linear operators from an algebraic standpoint', Glasgow Math. J. 8 (1967), 41-9.

120. F. F. Bonsall, Hermitian operators on Banach spaces, Colloquia

Mathematica Societatis János Bolyai 5. Hilbert space operators, Tihany (Hungary), 1970, 65-75.

121. F. F. Bonsall, 'An inclusion theorem for the matrix and essential ranges of operators', J. London Math. Soc. (to appear).

122. F. F. Bonsall and M. J. Crabb, 'The spectral radius of a Hermitian element of a Banach algebra', Bull. London Math. Soc. 2 (1970), 178-80.

123. R. Bouldin, 'The numerical range of a product', J. Math. Anal. Appl. 32 (1970), 459-67.

124. R. Bouldin, 'The numerical range of a product, II', J. Math. Anal. Appl. 33 (1971), 212-19.

125. A. Browder, States, numerical ranges, etc., Proceedings of the Brown informal analysis seminar, Summer, 1969.

126. A. Browder, 'On Bernstein's inequality and the norm of Hermitian operators', Amer. Math. Monthly, 78 (1971), 871-3.

127. A. Brown and C. Pearcy, 'Structure of commutators of operators', Ann. Math. 82 (1965), 112-27.

128. R. B. Burckel, 'A simpler proof of the commutative Glickfeld-Berkson theorem', J. London Math. Soc. (2) 2 (1970), 403-4.

129. B. Calvert and K. Gustafson, 'Multiplicative perturbation of non-linear m-accretive operators', J. Functional Analysis, 10 (1972), 149-58.

130. M. J. Crabb, 'The numerical range of an operator', Ph.D. thesis, Edinburgh, 1969.

131. M. J. Crabb, 'Some results on the numerical range of an operator', J. London Math. Soc. (2) 2 (1970), 741-5.

132. M. J. Crabb, 'The power inequality on normed spaces', Proc. Edinburgh Math. Soc. (2) 17 (1971), 237-40.

133. M. J. Crabb, 'The powers of an operator of numerical radius one', Michigan Math. J. 18 (1971), 253-6.

134. M. J. Crabb, J. Duncan, and C. M. McGregor, 'Some extremal problems in the theory of numerical ranges', Acta Math. 128 (1972), 123-42.

135. M. J. Crabb, J. Duncan, and C. M. McGregor, 'Mapping theorems and the numerical radius', Proc. London Math. Soc. (3) 25 (1972), 486-502.

136. M. J. Crabb and A. M. Sinclair, 'On the boundary of the numerical range', Bull. London Math. Soc. 4 (1972), 17-19.

137. C. Davis, 'The Toeplitz-Hausdorff theorem explained', Canadian Math. Bull. 14 (1971), 245-6.

138. N. P. Dekker, 'Joint numerical range and joint spectrum of Hilbert space operators', Ph.D. thesis, Amsterdam, 1969.

139. J. Dixmier, Les algèbres d'opérateurs dans l'espace Hilbertien, Gauthier-Villars, Paris, 1969.

140. J. Dixmier, Les C*-algèbres et leurs representations, Gauthier-Villars, Paris, 1964.

141. W. F. Donoghue, 'On the numerical range of a bounded operator', Michigan Math. J. 4 (1957), 261-3.

142. R. G. Douglas, 'On the operator equation $S^*XT = X$ and related topics', Acta Sci. Math. (Szeged), 30 (1969), 19-32.

143. H. R. Dowson, 'On the algebra generated by a Hermitian operator', Proc. Edinburgh Math. Soc. (2) 18 (1972), 89-91.

144. H. R. Dowson, 'A commutativity theorem for pre-spectral operators', (to appear in Illinois J. Math.).

145. R. J. Duffin and A. C. Shaeffer, 'Some inequalities concerning functions of exponential type', Bull. Amer. Math. Soc. 43 (1937), 554-6.

146. N. Dunford, 'Spectral operators', Pacific J. Math. 4 (1954), 321-54.

147. N. Dunford, 'A survey of the theory of spectral operators', Bull. Amer. Math. Soc. 64 (1958), 217-74.

148. N. Dunford and J. T. Schwartz, 'Linear operators, Part I', Interscience, New York, 1958.

149. M. R. Embry, 'The numerical range of an operator', Pac. J. Math. 32 (1970), 647-50.

150. M. R. Embry, 'Classifying special operators by means of subsets associated with the numerical range', Pac. J. Math. 38 (1971), 61-5.

151. P. A. Fillmore, J. G. Stampfli, and J. P. Williams, 'On the essential numerical range, the essential spectrum, and a problem of Halmos', Acta Sci. Math. (Szeged) (to appear).

152. S. Fisher, 'The convex hull of the finite Blaschke products', Bull. Amer. Math. Soc. 74 (1968), 1128-9.

153. W. Givens, 'Fields of values of a matrix', Proc. Amer. Math. Soc. 3 (1952), 206-9.

154. D. Gries, 'Characterizations of certain classes of norms', Numer. Math. 10 (1967), 30-41.

155. K. Gustafson, 'The Toeplitz-Hausdorff theorem for linear operators', Proc. Amer. Math. Soc. 25 (1970), 203-4.

156. P. R. Halmos, 'Numerical ranges and normal dilations', Acta Sci. Math. (Szeged), 25 (1964), 1-5.

157. P. R. Halmos, Math. Reviews, 41 (1971), 1368.

158. L. A. Harris, 'Schwarz's lemma in normed linear spaces', Proc. Nat. Acad. Sci. 62 (1969), 1014-17.

159. L. A. Harris, 'The numerical range of holomorphic functions', American J. Math. 93 (1971), 1005-19.

160. L. A. Harris, 'Banach algebras with involution and Möbius transformations', J. Functional Analysis, 11 (1972), 1-16.

161. S. Hildebrandt, 'The closure of the numerical range of an operator as a spectral set', Comm. Pure Appl. Math. 17 (1964), 415-21.

162. S. Hildebrandt, 'Numerischer Wertebereich und normale Dilatationem', Acta Sci. Math. (Szeged), 26 (1965), 187-90.

163. S. Hildebrandt, 'Über den numerischen Wertebereich eines Operators', Math. Ann. 163 (1966), 230-47.

164. E. Hille and R. S. Phillips, 'Functional analysis and semigroups', Amer. Math. Soc. Coll. Publ., volume 31, Providence, 1957.

165. J. A. R. Holbrook, 'Multiplicative properties of the numerical radius in operator theory', J. f. d. Reine und Angewandte Math. 237 (1969), 116-25.

166. V. Istratescu, 'Some remarks on the spectra and numerical range', Comm. Math. Univ. Carolinae, 9 (1968), 527-31.

167. V. Istratescu, 'On maximum theorems for operator functions', Rev. Roum. Math. Pures et Appl. 14 (1969), 1025-9.

168. V. Istratescu, 'On some normaloid operators', Rev. Roum. Math. Pures et Appl. 14 (1969), 1289-93.

169. G. A. Johnson, 'Matrix ranges of operators on Hilbert spaces', Ph.D. thesis, Edinburgh, 1972.

170. S. Kakutani, 'A generalization of Brouwer's fixed point theorem', Duke Math. J. 8 (1941), 457-9.

171. D. Koehler, 'A note on some operator theory in certain semi-inner-product spaces', Proc. Amer. Math. Soc. 30 (1971), 363-6.

172. D. O. Koehler and P. Rosenthal, 'On isometries of normed linear spaces', Studia Math. 36 (1970), 213-16.

173. B. Kritt, 'Generalized pseudo-Hermitian operators', Proc. Amer. Math. Soc. 30 (1971), 343-8.

174. B. Ya. Levin, 'Distribution of zeros of entire functions', Amer. Math. Soc. Translations, Providence, 1964.

175. G. Lumer, 'Bounded groups and a theorem of Gelfand', Revista Un. Mat. Argentina, 25 (1971), 239-45.

176. G. Lumer, 'Complex methods and the estimations of operator norms and spectra from real numerical ranges', J. Functional Analysis, 10 (1972), 482-95.

177. Yu. I. Lyubič, 'Almost periodic functions in the spectral analysis of operators', Dokl. Akad. Nauk, 132 (1960), 518-20.

178. Yu. I. Lyubič, 'On conditions for complete systems of eigenvectors of correct operators', Usp. Mat. Nauk, 18 (1963), 165-71.

179. Yu. I. Lyubič, 'Conservative operators', Usp. Mat. Nauk, 20 (1965), 221-5.

180. C. M. McGregor, 'Some results in the theory of numerical ranges', Ph.D. thesis, Aberdeen, 1971.

181. C. M. McGregor, 'Finite dimensional normed linear spaces with numerical index 1', J. London Math. Soc. (2) 3 (1971), 717-21.

182. C. M. McGregor, 'Operator norms determined by their numerical ranges', Proc. Edinburgh Math. Soc. (2) 17 (1971), 249-55.

183. C. H. Meng, 'A condition that a normal operator has a closed numerical range', Proc. Amer. Math. Soc. 8 (1957), 85-8.

184. C. H. Meng, 'On the numerical range of an operator', Proc. Amer. Math. Soc. 14 (1963), 167-71.

185. R. T. Moore, 'Adjoints, numerical ranges and spectra of operators on locally convex spaces', Bull. Amer. Math. Soc. 75 (1969), 85-90.

186. R. T. Moore, 'Hermitian functionals on B-algebras and duality characterizations of C*-algebras', Trans. Amer. Math. Soc. 162 (1971), 253-66.

187. I. Namioka and E. Asplund, 'A geometric proof of Ryll-Nardzewski's fixed point theorem', Bull. Amer. Math. Soc. 73 (1967), 443-45.

188. G. Orland, 'On a class of operators', Proc. Amer. Math. Soc. 15 (1964), 75-80.

189. T. W. Palmer, 'Characterizations of C*-algebras, II', Trans. Amer. Math. Soc. 148 (1970), 577-88.

190. T. W. Palmer, 'Real C*-algebras', Pacific J. Math. 35 (1970), 195-204.

191. T. W. Palmer, 'The Gelfand-Naimark pseudo-norm on Banach *-algebras', J. London Math. Soc. (2) 3 (1971), 59-66.

192. V. P. Potapov, 'The multiplicative structure of J-contractive matrix functions', Amer. Math. Soc. Translations, 15 (2) (1960), 131-243.

193. V. Pták, 'On the spectral radius in Banach algebras with involution', Bull. London Math. Soc. 2 (1970), 327-34.

194. V. Pták, 'Banach algebras with involution', Manu. Math. 6 (1972), 245-90.

195. H. Radjavi, 'Structure of A*A - AA*', J. Math. Mech. 16 (1966), 19-26.

196. R. Raghavendram, 'Toeplitz-Hausdorff theorem on numerical ranges', Proc. Amer. Math. Soc. 20 (1969), 284-5.

197. G-C. Rota, 'On models for linear operators', Comm. Pure Appl. Math. 13 (1960), 469-72.

198. B. Schmidt, 'Spektrum, numerischer Wertebereich und ihre Maximumprinzipien in Banachalgebren', Manuscripta Math. 2 (1970), 191-202.

199. M. Schreiber, 'Numerical range and spectral sets', Michigan Math. J. 10 (1963), 283-8.

200. I. Segal, 'Decompositions of operator algebras I, II', Memoirs Amer. Math. Soc. 9, 1951.

201. B. Sims, 'A characterization of Banach-star-algebras by numerical range', Bull. Australian Math. Soc. 4 (1971), 193-200.

202. A. M. Sinclair, 'Eigenvalues in the boundary of the numerical range', Pac. J. Math. 35 (1970), 231-4.

203. A. M. Sinclair, 'The states of a Banach algebra generate the dual', Proc. Edinburgh Math. Soc. (2) 17 (1971), 341-4.

204. A. M. Sinclair, 'The Banach algebra generated by a Hermitian operator', Proc. London Math. Soc. (3) 24 (1972), 681-91.

205. R. Sine, 'On a paper of Phelps', Proc. Amer. Math. Soc. 18 (1967), 484-6.

206. R. Sine, 'A note on rays at the identity operator', Proc. Amer. Math. Soc. 23 (1969), 546-7.

207. P. G. Spain, 'On commutative V*-algebras', Proc. Edinburgh Math. Soc. (2) 17 (1970), 173-80.

208. P. G. Spain, 'On commutative V*-algebras, II', Glasgow Math. J. 13 (1972), 129-34.

209. P. G. Spain, 'V*-algebras with weakly compact unit sphere', J. London Math. Soc. (2) 4 (1971), 62-4.

210. I. N. Spatz, 'Smooth Banach algebras', Proc. Amer. Math. Soc. 22 (1969), 328-9.

211. J. G. Stampfli, 'Adjoint Abelian operators on Banach spaces', Can. J. Math. 21 (1969), 505-12.

212. W. F. Stinespring, 'Positive functions on C*-algebras', Proc. Amer. Math. Soc. 6 (1955), 211-16.

213. K. W. Tam, 'Isometries of certain function spaces', Pac. J. Math. 31 (1969), 233-46.

214. E. Torrance, 'Adjoints of operators on Banach spaces', Ph. D. thesis, Illinois, 1968.

215. E. Torrance, 'Maximal C*-algebras of a Banach algebra', Proc. Amer. Math. Soc. 25 (1970), 622-4.

216. N. Th. Varopoulos, 'Tensor algebras and harmonic analysis', Acta Math. 119 (1967), 51-112.

217. H. Vogg,
Ph. D. thesis, Munich, 1971.

218. W. G. Vogt, M. M. Eisen, and G. R. Buis, 'Contraction groups and equivalent norms', Nagoya Math. J. 34 (1969), 149-51.

219. J. P. Williams, 'Similarity and the numerical range', J. Math. Anal. Appl. 26 (1969), 307-14.

220. J. P. Williams, 'Operators similar to their adjoints', Proc. Amer. Math. Soc. 20 (1969), 121-3.

221. J. P. Williams, 'On commutativity and the numerical range in Banach algebras', J. Functional Analysis, 10 (1972), 326-9.

222. Chr. Zenger and E. Deutsch, 'Inclusion domains for the eigenvalues of stochastic matrices', Num. Math. 18 (1971), 182-92.

223. T. Furuta and R. Nakamoto, 'On tensor products of operators', Proc. Japan Acad. 45 (1969), 680-5.

224. G. Lumer, 'Bochner's theorem, states, and the Fourier transforms of measures', (to appear).

225. I. S. Murphy, 'A note on Hermitian elements of a Banach algebra', J. London Math. Soc. (to appear).

226. H. Schneider and R. E. L. Turner, 'Matrices Hermitian for an absolute norm', (to appear).

Index

PB-8480-4
5-21

PB-8480-4
5-21